People of the Lake

Man; his origins, nature and future

Richard Leakey and Roger Lewin

People of the Lake

Man; his origins, nature and future

COLLINS *St James's Place, London*

First published 1979

© Richard Leakey and Roger Lewin 1979

ISBN 0 00 219502 X

Printed and bound in Great Britain by
Morrison & Gibb Ltd, London and Edinburgh

Contents

List of Plates

Introduction

by Roger Lewin

The autumn of 1978 marked the end of the first complete decade of fossil hunting at Lake Turkana. A more fruitful
and exciting 10 years in the search for human origins could hardly
be imagined: fragments of fossilised bone from several hundred
pre-human creatures have been unearthed from the dry
deposits on the lake's eastern shore and they slot together to
create a unique pattern outlining the complexity and direction
of human prehistory over the past two to three million years.
The past decade has witnessed the birth of a new view of the
dynamics of human origins, and, unquestionably, the discoveries made by the large team of workers at the Koobi Fora
camp have played the role of chief midwife at the birth.

That Richard Leakey should have been the leader of the
highly successful and productive Koobi Fora project is perhaps
inevitable – he is the son of the world-famous prehistorians
Louis and Mary Leakey, and he has much of the drive and
determination that his father deployed with such effect. And
yet the first firm decision Richard remembers making was that
he would not follow in his parents' footsteps!

Those footsteps blazed an astonishing pioneering trail in the
search for human origins in East Africa, a search that stretches
back through more than 40 years of eye-catching activity and
is focused principally at Olduvai Gorge on the edge of the
Serengeti Plain in Tanzania. It was at the Gorge that Mary
discovered in 1959 the shattered fragments of *Zinjanthropus*, a
robust man-like creature who lived almost two million years
ago. The event put East Africa on the palaeoanthropological
map which until then had been dominated by discoveries in
the limestone caves of South Africa.

Within a few years of the appearance of *Zinjanthropus*,
Olduvai Gorge yielded its second major hominid. This, Louis
decided, was even more closely related to direct human
ancestry: the hominid came to be named *Homo habilis*, the class

9

to which another famous skull, to be named 1470, discovered at Lake Turkana in 1972 also belongs. The discovery of *Homo habilis* at Olduvai Gorge, and the subsequent finds of even better specimens at Lake Turkana, provided the major thrust for a rethink of human origins.

It was actually yet another Leakey who first came across the Olduvai *Homo habilis*: the finder was Richard's elder brother, Jonathan, still a teenager at the time. It was a Leakey family tradition that all three brothers, Jonathan, Richard, and Philip, the youngest, accompanied their parents on field trips right from the moment they were born. Inevitably all three sons absorbed the basics of fossil hunting and became intimately acquainted with the lore of the countryside – as well as picking up doses of bilharzia and malaria on the way!

Jonathan now virtually never indulges in fossil hunting; Philip occasionally joins the efforts of the team at Olduvai and nearby Laetoli; leaving Richard as the principal heir to his father's charisma since the great man died in October 1972.

Richard's initial reluctance to be a fossil hunter slowly eroded as, sporadically at first, he joined small expeditions in various parts of Kenya – nothing spectacular came from them, but it was enough to whet his appetite. Meanwhile he had established his independence, both psychological and financial, by setting up a photographic safari business, an enterprise that had left him little time or motivation for pursuing a university education. Then, in 1967, at the age of 22, he led his first major expedition, the Kenyan contingent of an international study of sites along the lower Omo River in Ethiopia. This was a decisive step because it was while he was in transit for the expedition that he spotted the potential of the eastern shore of Lake Turkana (or Rudolf as it was then known).

Urgent business had forced Richard temporarily to leave the camp and return to Nairobi. On the return journey in the small two-seater plane the pilot decided to fly along the eastern shore of Lake Turkana rather than going over the more normal western route so that he would avoid the threat of local bad weather there. Glancing from the window of the passenger seat Richard saw what he thought might be eroded layers of lake deposits on the eastern shore – this would be a promising site for finding fossils, he mused.

And so it was. A reconnaissance trip by helicopter hired

especially for the occasion from the US party confirmed that the inhospitable-looking black rocks were in fact fossil-bearing sandstone layers and not lava flows as everyone had always imagined. With a small grant from the National Geographic Society, and in the company of five colleagues, Richard, still lacking the conventional academic credentials, set off in 1968 for the first survey expedition of the lake deposits. (The grant, incidentally, was given to Richard against Louis' advice, though Richard suspects that his father may have adopted the negative position as a way of encouraging his son, such was the complicated relationship between them at the time.)

During the next few years a camp was set up at Koobi Fora on the shores of Lake Turkana. The 500-mile journey from Nairobi was sometimes covered by a gruelling, though spectacularly beautiful, drive direct to the camp, and on others by an easier 12-hour road stint to the lodge on the lake's west shore, the last leg of the journey being completed by boat. And on one occasion the team drove to Marsabit some 200 miles south of Koobi Fora where they picked up some camels which they used as pack and (not very efficient) riding animals.

In the most meagre conditions, and against difficult odds which at one time included having all their equipment destroyed by the local people, the determined team built a permanent base. Located on a spit jutting out into the lake, the camp, which consists of a group of stone-built thatched bandas, can now accommodate 50 people comfortably – and 70 uncomfortably! Being surrounded on three sides by the lake helps cool the camp environs by as much as 10°F: in January and February the temperature out in the field can reach 115°F day after relentless day, and even in the cooler months of the year exploration is done routinely with the thermometer hovering around 100°F. The relief of returning to a considerably cooler camp with the opportunity for a bathe in the lake need not be described!

Each year dozens of workers from many countries in the world come together as a team for a field research season. Richard, whose principal job is director of the National Museums of Kenya, visits Koobi Fora periodically, as doe Meave, his wife, and their two children Louise and Samira. Richard now does the trip in two hours and 40 minutes, piloting his own single-engined Cessna. A highly efficient fossil-hunting team, led by Kamoya Kimeau, is based perma-

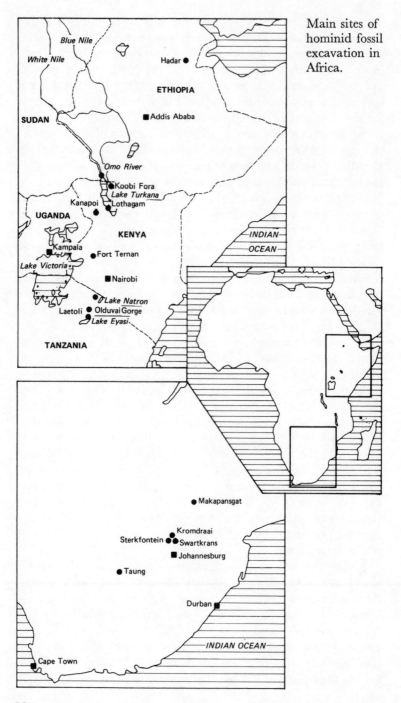

Main sites of
hominid fossil
excavation in
Africa.

Blue Nile

White Nile

Hadar ●

ETHIOPIA

SUDAN

■ Addis Ababa

Omo River

● Koobi Fora
Lake Turkana

Kanapoi ● Lothagam

UGANDA

KENYA

INDIAN
OCEAN

Kampala ■

● Fort Ternan

Lake Victoria

■ Nairobi

● Lake Natron

Laetoli ● Olduvai Gorge
Lake Eyasi

TANZANIA

● Makapansgat

Kromdraai
Sterkfontein ●● Swartkrans

■ Johannesburg

● Taung

Durban ■

INDIAN OCEAN

■ Cape Town

12

nently at the lake, but whenever an important hominid fossil is discovered Richard insists on excavating it himself. The success of the camp is undoubtedly its commitment to team effort, with contributions of many different types coming from many different sources.

Although some fossil finds from Koobi Fora have received more publicity than others, the special contribution of the site to an understanding of human origins derives from the total picture that the fossil collection presents as a whole; this will be elaborated upon in the rest of the book. In attempting to reconstruct the lives of our long-dead ancestors from the fossil bones and shaped stones they left behind in the fossil record it is impossible to separate completely one activity from another: hunting impacts on social organisation, which in turn has implications for the nature of language, which in its turn affects the context in which stone tools are constructed. In the march of mankind along the long evolutionary road many factors interacted with each other in an integrated matrix.

In telling the story of human origins we attempt to tease out of that matrix some of the most important factors, while being aware of the danger of over-simplification. And we try to display the intricately interacting processes, while avoiding total confusion!

In chapters 2 to 5 we concentrate on the bare bones of human evolution, what the recent fossil discoveries tell us about biological structure of our origins. We can learn something about basic economy and social organisation by looking at technologically primitive gathering and hunting people (chapter 6) and speculate about the origins of these characteristics in our ancestors (chapter 7).

Intelligence and language, and their inextricable social and cultural context, form chapters 8 and 9, while the thorny issue of sex and sex roles comes in the next chapter. Last of all we tackle the inescapably political aspect of human prehistory: are humans innately aggressive? Is war and bloody oppression an inevitable element of human history?

Main sites of hominid fossil excavation in Africa and Eurasia.

CHAPTER 1

People of the Lake

The scene is Africa, 15 million years ago, in the era we call the Miocene (it spans 20 million to five million years ago). It is only a few million years since the continent collided once again with Eurasia, a collision that was powered by the movement of vast plates that make up the surface of the earth and upon which the continents literally float like huge granite boats. At the time the African landscape was very different from today's: the highlands of Kenya and Ethiopia did not exist, and the carpet of forest that today stops west of the Rift Valley, in Miocene times swept on to the Indian Ocean. Forest and woodland were the predominant environments through West and East Africa. But the geological forces that had rammed the continent into Eurasia were still stirring, and they were about to start tearing the continent apart, along the Great Rift Valley.

Although the movement of the massive plates that is gradually widening the Rift Valley is not dramatic (it is opening up by about one millimetre a year compared with the one centimetre or so that is added to the distance between Europe and the US each year), the geology is made sensational by the growth of two massive blisters, one in Kenya, the other in Ethiopia. Swellings of molten lava from deep down in the earth's mantle heaved up the land by 3000 feet and more to form these two nations' highlands. The crust groaned under tremendous pressure as it was forced higher and higher: in the end the strain was just too much; the crust had to crack, and it did. Countless trillions of tons of rock crashed downwards as fault lines opened up, stretching from northeast to southwest: the blisters were lanced, creating the first visible signs of what was to become the Great Rift Valley.

Inevitably the boiling, swirling lava and gases found ways to escape, tearing explosion craters in the hillsides and building huge volcanoes that, in middle Miocene times (say 12 million years ago), soared up 20,000 feet. They would have been snow-covered then, just as Mounts Kenya and Kilimanjaro are today. Through the ages the ancient volcanoes were ground down by the elements, leaving mere stubs a tenth of the original size; their pulverised rocks contributed to the layer upon layer of deep sediments in the lava floor of the valley, a floor that is now more than 4000 feet thick. Through the ages too that floor sank lower and lower, partly because of the weight of the sediments and lava and partly because of the movement apart of the plates. These days, in the region of Lake Baringo, which was the highest point of the Kenyan blister, the walls of the valley rise 3000 feet from the lava floor, and they are virtually sheer. And the bottom of the valley, through Tanzania, Kenya, and Ethiopia, is littered with lakes and more or less extinct volcanic cones. It is littered too with fossil sites rich in the remains of our early ancestors and their australopithecine cousins.

As a geological spectacle the Rift Valley of today is un-paralleled. And as a major cause of perturbation of climate in the past it was equally stunning. As the blisters rose higher and higher they eventually reached a point at which they threw the land to their east into rain shadow, thus depriving the thirsty tropical forest there of its sustenance. The forests of East Africa shrank, producing a patch-work of embryonic savanna (open terrain) and woodland, and leaving a scattering of West African-like trees, birds, butterflies and animals in the Arabuko Soreke coastal forest just north of Mombasa and up to the Tana River as a reminder of times past.

The pruning of East Africa's forests by the birth of the highlands is ecological modification on a grand scale. But the cracking and blistering geological face of this area generated more subtle, local effects too: within a radius of just a few miles around any part of the rift, but particularly in the highest areas, there is a mosaic of dense tropical forest, semi-arid desert, alpine meadows, grasslands, open woodland, and every eco-logical shade in between. It is a remarkable piece of topography, and it has been that way since the middle Miocene right up to the present. The creation of this matrix of ecologies as the blisters swelled and cracked provided an unusual diversity of habitats for the animals there to exploit. And this may well have been an important factor in speeding along the pace of hominid evolution in East Africa. (Hominid is the name used for humans and their close evolutionary relatives such as the australopithecines.)

As the Rift Valley sweeps northwards out of Kenya and into Ethiopia in modern-day Africa it forms the spectacular Lake Turkana basin. Spectacular, not only for the stunning beauty of the lake itself and its powerfully stark surroundings, but also for its rich treasure trove of pre-human fossils buried in the layered deposits on the eastern shore of the lake. Beginning with a small tentative expedition in 1968, it is here that Richard Leakey leads the search for ancient human ancestors in Kenya.

The long shallow waters of the lake, which stretch 155 miles north to south and up to 35 miles east to west, sparkle green in the tropical sun: someone called it the Jade Sea, a very apt name. At the south a barrier of small volcanic hills prevents the lake spreading further down into the arid lands of northern Kenya. From the west side rises the Rift Valley wall, a range of mountains with some peaks of more than 5000 feet. This is the land of the Turkana people, a tall, elegant, pastoralist tribe. Beyond are the mountains and forests of Uganda.

Pouring its silt-laden waters into the north end of the lake is the River Omo, a huge river that drains the Ethiopian highlands to the north and meanders tortuously as it nears its end at the border with Kenya where it reaches the Jade Sea. Fly over Lake Turkana and you see the orange waters of the Omo carried miles by their massive momentum into the green lake until they finally disperse, orange into green—a spectacular sight. As the river reaches the lake the sudden barrier to its progress forces it to dump its burden of silt, so creating an enormous delta. This process has been going on for at least four million years, and it is the ancient river and lake sediments that have helped preserve the fossils we find today.

One important source of fossils is the lower Omo valley itself, where over a period of about four million years more than 3000 feet of sediments built up, trapping the skeletons of ancient hominids and examples of their unusual stone technology. But the eastern shore of the lake is an even richer fossil treasure trove.

The Koobi Fora Research Project has its camp slightly more than half way up the eastern shore, on a spit of land known as Koobi Fora. It is a good place to camp, not only because the widespread fossil sites are relatively easily reached from there, but also because it offers a welcome and refreshing bathe after a scorching day in the field. So far the local crocodiles have contented themselves with simply watching the spectacle of wallowing archaeologists, taphonomists, palaeontologists, and the like, and have not been tempted to add any of them to their menu!

Today the lake has no outlet, except for evaporation: its resultant alkalinity encourages the growth of certain algae which are the favourite food of pink flamingoes – this is a common feature of the many Rift Valley lakes. But Lake Turkana has not always been like this. For instance, 10,000 years ago the waters stood a staggering 200 feet higher than today. The lake must have been gargantuan then, expanding principally up the Omo valley, further south into Kenya, and east toward the highlands that now rise 20 miles from the lake margin. When the water level was so much higher the lake did have an outlet: it was a source of the Nile; the presence of enormous Nile perch and Nile crocodiles tells us that.

We don't know why, 10,000 years ago, the lake was so high. Nor do we know why it plummeted so dramatically. Local climatic aberrations might have been the cause. Perhaps enormous earth movements in this geologically unstable area allowed the waters to drain away or changed the fate of systems bringing the rainfall from the surrounding hills. Probably both factors played their part. In any case, Lake Turkana has been an extremely active body of water over the past four million years or so, with at least three major changes in level (though probably not quite so dramatic as that 10,000 years ago), interspersed with more or less gentle fluctuations. Currently the lake is shrinking, and the walk for the evening bathe is at least 20 yards longer than it was in 1969 when the camp was established.

A restless lake of this sort, fed not only by a major river such as the Omo, but also by numerous seasonal streams and rivers, is ideal for creating a fossil record: the bones may be rapidly entombed in fine silt, a process that first of all prevents the bones decaying, and secondly encourages the gradual replacement of the bone's own chemicals by hard rock minerals. Basically there are three circumstances around the lake where these essential processes occur: first, at the lake shore itself where gentle lapping waves throw a coat of silt over bones lying in the shallows; second, in the beds of streams tumbling down to the lake's edge; and third, between these extremes, at the point where the stream nears the lake water, thus forcing it to drop its silt.

For anyone wanting to become part of some future fossil record, this last situation would be the one to choose in which to expire. Certainly, it is the one that has yielded most of the hominid fossils at Lake Turkana.

From a fossil-hunter's point of view, preservation and mineralisation of bones is just the first step in putting together a record of the past. Equally important is being able to get at that record. What do you do if you suspect that fossils may be buried in superb preserving deposits 100 feet below where you are standing? Nothing, because the chances of actually finding anything by such a hit and miss method are virtually zero. Not to mention the enormous effort that would be involved. In any case, most deposits would demand a major mining expedition.

A few mental visits to parts of the Rift Valley, and elsewhere, make this point. The deposits of an ancient, and now vanished, lake at Olduvai in Tanzania, are 300 feet thick; fortunately a recent seasonal river has sliced its way down through them, exposing in the 25-mile-long gorge a vertical record of the past. At the Hadar site near the Awash River in northern Ethiopia, streams and wind have cut through 600 feet of prehistoric lake sediments. Further south in Ethiopia, the Omo River laid down more than 3000 feet of sediments over a period of four million years. Here the restless earth has recently tipped the sediments, thus exposing the previously buried layers, with the oldest in the east near to the present river and leading to the youngest in the west. Perhaps the most breathtaking example, though, comes from the Siwalik hills in Pakistan where British palaeoanthropologist David Pilbeam is working. There, during the past 15 million years, countless millions of tons of silt have poured off the Himalayas, accumulating below in a layer cake of time more than four miles thick! Pilbeam, who currently teaches at Yale, is lucky because rivers and streams have done most of his initial excavating for him. And at Koobi Fora, the previous ages encapsulated in a modest 100 feet of sediments have been opened up to us partly by their recent tilting (as at the Omo) and partly through their erosion by wind and streams.

In all these places you prospect for fossils by walking around looking for signs of them on the surface: a glint of bone may be the only visible trace of a complete skull buried just under the surface. Or it may merely be a single disappointing fragment. With every season's rains new fossils may be exposed. But if they lie on the surface too long they rapidly disintegrate, to be lost forever.

The earth movements that are characteristic of the Rift Valley's nature were crucial to the formation of a string of ancient lakes through Ethiopia, Kenya, and Tanzania, lakes that both must have been attractive as places to live for our early ancestors and their relatives, and were therefore convenient places in which to die on the way to becoming part of the fossil record. (We are, of course, not ascribing altruistic motives to the early hominids in their chosen locations for expiring!) The continued geological stirrings have helped give us a glimpse of the past by heaving some of the sediments upwards, thus exposing them to the ravages of erosion and the inquisitive attentions of modern archaeologists.

Suppose, now, we are back on the eastern shores of Lake Turkana, a few miles north of the Koobi Fora spit, two and a half million years ago. What might we see? Standing by the shores we would be aware of crocodiles basking in the tropical heat on sand spits pointing finger-like into the shallow waters. Hippos wallow, occasionally exploding watery sighs and making waves as they jostle each other lazily. The air is punctuated by the slap of wings as a group of pelicans take noisily to the air, squawking crossly for reasons best known to themselves.

A little more than five miles away to the east, savanna-covered hills rise up from the lake basin, sliced here and there by forest-filled valleys. At one point the hills are breached by what is obviously a large river that has snaked its way down from the Ethiopian mountains. We can't see the river because its path is followed by a lush growth of trees and bushes: wild figs, acacia, and Celtis grow thickly. As the river reaches the floodplain of the lake it shatters into a delta of countless streams, some small, some large, but each fringed by an attentive line of trees and bushes.

As we walk up one of the stream beds – dry now because there have been no rains for months – we might hear the rustle of a pig in search of roots and vegetation in the undergrowth. As the tree cover thickens we catch a glimpse of a colobus monkey retreating through the tree tops. Lower down, mangabeys feed on the ripening figs. In the seclusion of the surrounding bushes small groups of impala and water buck move cautiously. By climbing a tree we could see out into the open where herds of gazelle graze and troops of gelada-like baboons forage in the grass and under stones and bushes.

After going about a mile up the stream we come across a scene that is strangely familiar but which nevertheless we have never observed before: a group of about eight creatures – definitely human-like, but definitely not truly human – are before us, some on the stream bed, some on its sandy bank. Two adult females are making piles of roots and nuts; they are emptying what appear to be containers made from animal skins. Another adult, a male, has just finished digging a hole in the stream bed, partly with his hands, partly with a stick. Children crowd round him, going down on their hands and knees to scoop up the water his excavation has reached. He shoos them away, and then fills a folded leaf with the cool water which he then gives to another adult male who is lying on the bank – he looks ill.

The scene is a mixture of industry and leisure: children play, some digging in the stream as the big male had just done, others inexpertly knocking two stones together at the feet of an adult who makes a simple tool with ease, and others just have fun chasing through the bushes.

Suddenly there is a shout – at least it sounds like a shout. Everyone turns in the direction of the call to see a group of adults, mostly males, walking excitedly towards the camp. They are carrying hunks of hippo flesh, and they are obviously pleased with themselves. They had been wandering along a tree-lined stream bed about a mile south of their camp early that morning and had stumbled across the freshly dead animal. So, after collecting some lava cobbles from some distance away towards the hills, they made some cutting tools and proceeded to slice off generous pieces of meat. And when they had eaten some of the tasty liver in celebration, they carried the meat triumphantly to the camp. The departure for the camp came none too soon as the steadily growing numbers of hyaenas were rapidly losing patience at being kept away from a meal they clearly thought was rightfully theirs.

The hippo was so big that it made good sense to butcher the carcass in this way rather than stagger back to the camp with the meat still clinging to a heavy leg bone, something they would have had no hesitation in doing with, say, a gazelle.

21

Two females and a youth who were just about to set off carrying skin containers and newly sharpened digging sticks in search of roots, berries, and nuts, changed their minds and stayed for the feast. It is a feast in which everyone joins, the meat being sliced up with razor-sharp stone flakes by the males who found the animal. A latecomer, a male who comes down the stream holding a bunch of roots in one hand and a dead hare in the other, also gets a share.

Compared with the tranquillity of just a little while ago, the camp scene is now alive with the hubbub, excitement and some squabbling of eating meat: it does not happen every day, and they clearly enjoy it. Judging by the variety of noises they are making and the responsive interactions, we can say that they are communicating with each other. They touch each other a lot too.

If we chose to stay on, we would see that some days later the group would move on, one of the older males helping his sick companion, for he is still weak. Behind they leave their stream bed campsite littered with broken stones, tools, flakes, old digging sticks, splintered nut shells, and one or two moulding uneaten tubers. They go in search of another sandy place that is also free from the vicious spike grass that thrives on the lake's floodplain; and they will want the shelter of trees again too, not just to escape from the glare of the sun, but also to flee into if they are threatened by predators. Probably they will camp in another dry stream bed. And perhaps they might be lucky again and find another bonanza of meat as part of their primitive hunting and gathering economy.

These episodes are, of course, pure fantasy, but we construct them around as many facts and inspired guesses as we can. There *is* an ancient living site in the place we describe (it is called the KBS site, KB being the initials of Kay Behrensmeyer, the scientist who found the site). And the bones of a hippo *do* lie surrounded by stone tools about a mile to the south. Although the two sites are about the same age (somewhat more than two million years) we do not suggest that the occupants of the one really did butcher the animal at the second site. But as a reconstruction of lifestyle, the scenario stands as a valid view.

The rains must have come shortly after the hominids left their living site; and the level of the water course must have risen slowly and steadily. We know this because even tiny flakes of stone light enough to blow away in a strong wind remained to be buried by silt, together with the bigger flakes and stone tools – and a fig leaf! The leaf subsequently decayed, leaving only an impression in the fine deposits to remind us of the cool shade its parent tree must have thrown on our ancestors. A more fanciful interpretation is that perhaps, with the inexorable advance towards manhood, the innocence of the Garden of Eden stops here?

CHAPTER 2

In the beginning

To understand something about the creatures by the lake we have to go back a further 17 million years. Twenty million years ago, before the Rift Valley was more than distant subterranean rumblings, the African forest was the home of the ape. The Miocene was the Age of the Ape. Compared with today, when the African apes (chimpanzees and gorillas) are confined to just a handful of increasingly endangered habitats, the Miocene ancestors practically owned the place: not only were there a lot of individuals, there were a lot of different species too, and they exploited virtually every niche that the forest and woodlands had to offer – some lived gorilla-like lives, others made a living more like gibbons, and still others lived lives for which there is no contemporary model. These apes, one of which was eventually to give rise to the hominid family, were at the time the last word in the evolution of the primate order, the zoological order to which we belong, accompanied by apes, monkeys, and small nimble prosimians such as the long-legged sifaka, the big-eyed bushbaby, the diminutive potto, and the elegant lemur.

The primate order was born around 70 million years ago, at about the time of the demise of the amazing dinosaurs. It was a time of dramatic change in players on the earth's animal stage: at one point in time it was the dinosaurs who played all the leading parts, with a few small primitive mammals filling insignificant, submissive roles. Then, with what must be considered unseemly haste for the normally sedate procession of evolutionary advance, the dinosaurs vanished from the scene over a space of just a few million years, leaving the stage to be taken over by the mammals.

As the dinosaurs slid into their monumental decline, a small tree-shrew-like creature snuffled its way out of the thickly covered forest floor and took to the trees, eventually to become the first of the primates. For the next 30 million years the primate order proliferated, producing scores of tree-living, insect-eating, small nocturnal animals. Their millennia upon millennia of predation upon insects while suspended high on narrow, treacherous twigs equipped them, via the cutting edge of natural selection, with grasping fingers tipped with nails in place of clumsy paws with claws, and with front-facing rather than sideways-looking eyes. Throughout this time new primate species were getting bigger, something that frequently happens when a family is expanding in a bullish evolutionary environment. Many of them switched their activities from night to light: they became diurnal rather than nocturnal. And because of the enormous practical advantages of seeing in colour rather than monochrome, it was inevitable that some of the diurnal prosimians should eventually evolve the neural apparatus for appreciating the chromatic world around them.

So, even at this stage in our heritage, some 40 million years ago, apparatus that is so important to us as human beings now – grasping hands, stereoscopic vision, and the gift of seeing in colour – had already emerged. The real evolutionary secret of humanity has always been to keep its equipment simple and adaptable, not to become specialised into a biological dead end.

Just when the prosimians seemed settled as a highly successful group of tree-dwellers, they gave issue to the monkeys, who promptly ousted their ancestors and dominated daytime life aloft. With the monkeys came a change in primate menu, shifting from emphasis on insects to emphasis on plants (leaves and fruit). And they were bigger than the prosimians.

But the monkeys' ascendancy was not to last long, for about 30 million years ago came the apes, progeny of the monkey stock. Bigger than monkeys, the many different types of apes that evolved exploited a whole range of ecological opportunities, some learning the trick of hanging suspended by three limbs from branches, plucking otherwise inaccessible fruit with the fourth, while others actually ventured back down to terra firma. (Although the continent of South America at this time carried its evolving population of so-called New World monkeys as it drifted westward across what was to become the Atlantic Ocean, finally colliding with North America just a few million years ago, it never saw an indigenous ape: the monkeys there simply did not evolve along that biological path. Why, we do not know, but it means that the Americas were denied the opportunity of offering a home to newly emerging hominids; the continent had to wait until just a few tens of thousands of years ago before man set foot on its shore.)

Although apes and monkeys were and are relatively eclectic in their food tastes, they depended heavily on leaves and fruit, a dietary predilection that essentially confined them to the tropics: chimpanzees did not actually object to the cold of northern Europe, but what they could not cope with was the unappetising offering of pine needles as the only greenery on the bleak winter trees. The same was true for their Miocene ancestors.

Like the diminutive prosimians and the agile monkeys before them, the Miocene apes thrived and proliferated, but the ones in which we are most interested as the probable forerunners of man are the so-called woodland apes, or dryopithicenes as they are known. Things being what they are in the world of palaeontology, the proliferation of species in the Miocene apes is probably exceeded by the range of ideas of how they should be named. The source of the confusion is, inevitably, the sparsity of fossil evidence. Here we will talk about just three ape-like creatures that lived around 15 million to 10 million years ago: *Gigantopithecus*, *Sivapithecus*, and *Ramapithecus*.

The last of these, *Ramapithecus*, a small creature (perhaps close to three feet tall), is currently favourite as the first true hominid. *Sivapithecus* was bigger than *Ramapithecus*, but their cousin *Gigantopithecus* was probably enormous. All we have of *Gigantopithecus* so far is a few examples of its massive grinding jaws, and we know that as time passed this creature evolved bigger and bigger jaws, finally lumbering into extinction around half a million years ago with a stature that must have dwarfed even that of a modern gorilla.

Now, if we are absolutely honest, we have to admit that we know virtually nothing about *Ramapithecus*: we don't *know* what it looked like; we don't *know* what it did; and, naturally, we don't *know* how it did it. But with the aid of jaw and tooth fragments and one or two bits and pieces from arms and legs, all of which represent a couple of dozen individuals, we can make some guesses, more or less inspired.

First, the facts. The group of animals we call *Ramapithecus* were small, probably no bigger than three feet, and, judging by the kind of environments where their bones became fossilised, they were never far from trees. Rather than romping around in dense forest, it looks as though *Ramapithecus* preferred forest fringes and woodland. Specimens of our putative ancestors have turned up in many places in the Old World: Kenya, India, Pakistan, Greece, Hungary, Turkey, and China. This little ape-like animal was obviously well adapted to the ecological conditions of the time. For a start, the area bathed in a tropical climate stretched further north than it does now, encompassing southern Eurasia as well as Africa. And, as well as the 'local' environmental changes that rippled through East Africa between 10 and 15 million years ago as a result of the birth of the Rift Valley, there is also some indication of a general shrinkage of forests throughout the Old World at this time. Such a change almost certainly was influential on the evolutionary metamorphosis that the diminutive *Ramapithecus* was about to embark upon.

A young Yale student, Edward Lewis, was first to discover the putative hominid ancestor. In 1932 he was digging around Haritalyangar, a cluster of villages in the Siwalik Hills about 100 miles north of New Delhi. On his lone expedition, equipped with pack horse, Lewis came across a fossilised upper jaw of something that was definitely ape-like, but had a number of unusual characteristics that encouraged him to take the plunge and suggest that it belonged to the human family, Hominidae. Lewis named his specimen *Ramapithecus brevirostris*, *Ramapithecus* being the genus name and meaning Rama's ape,* and *brevirostris* the name of the species, meaning short-snouted.

It was basically the shortness of the face that persuaded Lewis to stick his neck out and name his find as the first hominid. But there was also a little hole in the jaw that added to his conviction: the hole had been left by the canine tooth, which, though nowhere to be found, was evidently significantly smaller than it should have been if its owner had been an ape.

This was definitely something different, an animal that was clearly exploring a new way of life. But the world wasn't ready for Lewis's discovery, and it lay ignored for more than 30 years. It was Louis Leakey, Richard's father. who found the Kenyan specimen of *Ramapithecus* in 1961 at a place called Fort Ternan in the south of the country. The first piece to be found was an upper jaw (a palate); some time later a closely fitting lower jaw turned up – they may well have belonged to the same individual. Louis Leakey found some splintered animal bones close

---

* The name of a prince in an Indian epic poem.

to the fossilised hominid that he thought had been broken deliberately with a blunt instrument. And he found a stone which, he claimed, showed signs of having been used as a 'bashing' instrument. Was this *Ramapithecus*, who had lived 14 million years ago, breaking the bones to get at the succulent marrow inside? Perhaps, but the evidence is really too tenuous to be anything like certain. Then, in the midst of rising scientific debate stimulated principally by Louis Leakey, Elwyn Simons and David Pilbeam, also at Yale, set themselves the task of examining and reclassifying the fossil apes of Miocene, and they not only agreed with Lewis, but they also classified as *Ramapithecus* some upper and lower jaw fragments that had come from India and West Pakistan and had been added to the Yale collection. During the subsequent ten years fossil hunters found many specimens of what looked to be an early hominid, but because people were usually unable to compare their specimens with the Yale material, there was a wondrous blossoming of names to describe what was basically the same type of animal: *Graecopithecus*, *Rudapithecus*, and *Kenyapithecus*, all were coined, but are now dropped in favour of Rama's ape.

The root stock of *Ramapithecus* was one of the earlier Miocene apes that had thrived in Africa before the continent collided with southern Eurasia around 16 to 18 million years ago, thereby creating the Mediterranean. The newly formed sea thereafter periodically dried out, leaving a collection of four or five huge 'puddles', interlaced with swamp and woodland, which were topped by the Atlantic from time to time in a series of cycles. Only within the last few million years did the Atlantic waters again break through between southern Spain and north Africa to form what is now the refuse dump of dozens of nations.

Because the Mediterranean did not bar its way around 15 million years ago, *Ramapithecus* was able to migrate directly into southern Europe; it must have taken him rather longer to have reached southern Asia. Because of this pattern of migration it comes as no surprise that the oldest *Ramapithecus* fossils found in Eurasia are in Turkey (around 15 million years), whereas those in Pakistan are much younger. During their colonisation of the new continents, and its continued occupation of home-land Africa, the basic *Ramapithecus* stock became widely dispersed, adapted to local conditions, and thus several different species emerged. Only one of these species eventually gave rise to the later hominids, a process that, as we will suggest, occurred only in Africa.

In tracing lines of descent, the shapes of the teeth are very important evidence in the fossil record. If you were able to persuade a modern gorilla to cooperate in a dental examination you would notice that its teeth have a very characteristic arrangement: they are set as on three sides of a rectangle, with the large protruding canines standing up like turrets at the corners of castle walls; the incisors are large; and the molars, which are marked by undulating cusps, are obviously designed for chewing. A gorilla's menu consists almost monotonously of leaves, stalks, and the odd succulent larva that might be clinging to the undersides of leaves. He uses his row of incisors, and sometimes the canines, to tear at the leaves and strip stalks, and then crushes the greenery between his molars with a front to back rocking motion of the jaw: the impressive protruding canines make side-to-side chewing impossible. Chimpanzees, which are rather more eclectic in their tastes, including fruits, nuts, and small animals in their varied diet, have very much a gorilla-like tooth pattern; its canines are slightly smaller than its cousin's, but they are still big enough to make side-to-side chewing an impossible proposition.

How does *Ramapithecus* compare with the modern apes? Its incisors are small; its molars are big and flattened with wear; its canines are small; and its jaw is V-shaped rather than rectangular; moreover, the lower jaw bone, the mandible, is strikingly robust. Most of these features are unmistakable signs of an animal that feeds on something tough that demands rigorous grinding. And, from marks on some specimens of teeth and the general architecture of the lower jaw, we know that when *Ramapithecus* sat chewing in the wooded Miocene landscape, its mandible moved from side to side, just as every self-respecting herbivore's jaw moves today.

Given its probable small body, *Ramapithecus*'s molars are relatively much bigger than either chimpanzees' or gorillas'. The same can be said for *Sivapithecus* and *Gigantopithecus*: their cheek teeth are relatively massive too. As well as this, the layer of white enamel on their teeth is much thicker than that of modern apes. Modern humans, incidentally, have a generous endowment of enamel, as did our ancestors in the Pliocene and Pleistocene. This commonality should alert us to

29

the special role that prehistoric menus played in our evolution and is absent in the modern apes!

Before we slip into a mood of total speculation, it is also worth trying to squeeze out of the miserable fragments of petrified limb bones some clues about how *Ramapithecus* got around, because our upright posture is another key to our evolution. A small fragment of lower arm bone from an eight million year old site in Pakistan is so slender as to imply that its owner would have been unwise to use it habitually for walking on. Does this belong to *Ramapithecus*, and, if so, does it mean that by this time *Ramapithecus* had abandoned the more 'primitive' quadrupedal walking (in the style of modern baboons perhaps) and taken to staggering about on their back legs? We simply cannot be certain, but it must have happened some time because by the time reasonable hominid fossils appear (about three million years ago) our ancestors were walking about with what almost certainly was a respectable upright gait.

A possible sequence for the early stages in the emergence of humanity from a forest ape, therefore, is this: first, between 15 and 12 million years ago changes in the environment opened up a new ecological niche for a woodland animal that could survive on tough food; second, somewhere between three and 12 million years ago, possibly around eight, there were important advantages in being able to walk around on two rather than four legs: these initial changes are then followed by more social and ecological adaptations, and the pace of evolution begins to warm up.

Why did *Ramapithecus* take to eating tough fibrous foods, a life-style that must have demanded more and more time on the ground rather than in the trees? Why did its canines shrink? Why did it start to walk around on two legs, when, by all accounts, walking on four is much less expensive energetically? These are the sort of questions to which we would like the answers, but to which, for the moment, we have only guesses.

If, as seems possible, the middle Miocene was a time when continent-wide forests began to shrink, then the apes living there would be squeezed in a vice of environmental stress: the scores of different forest dwellers would be pressed into ever-increasing confrontation with each other, competing for the diminishing resources – the cutting edge of natural selection would become very keen. But the loss of forests on the one hand meant the growth of woodland and savanna on the other. And this offered an opportunity to any animal willing to take it. Any open savanna would almost certainly already be colonised by efficient herbivores and carnivores that preyed on the grass-eater: competition here would have been keen for a naïve newcomer. Woodland would have been a much safer bet for an erstwhile forest dweller.

We are not suggesting, of course, that one day a group of *Ramapithecus*, decided that perhaps it was worthwhile venturing out into the more open woodland, and later perhaps even the savanna, to see if a living could be made. Evolution does not work like that. Nevertheless, opportunism is the hallmark of the higher primates, and it may be that it was a particularly generous endowment of opportunism that, in the proper biological context, encouraged *Ramapithecus* to leave the stress of the shrinking forests and to stake a claim elsewhere.

A less charitable interpretation is that the poor creature was squeezed out of the home it had known for so long by the intense competition and that it was forced to scratch out a living (literally) in a new and relatively impoverished ecological niche. Whatever the reason for the move into the new habitat, we can guess that *Ramapithecus* did not find itself overwhelmed by a superabundance of food. Instead, it must have spent much of its day searching for roots, seeds, stems, nuts, tough fruits, and insects, most of which had to be pulverised between efficient millstone-like teeth.

This kind of lifestyle is not unlike that now pursued by the hairy gelada baboons that live in the highland plateaux of Ethiopia. These animals spend their days shuffling along on their haunches turning over stones in search of roots, shoots and insects, and they too pulverise their food between flattened, millstone-like molars. They too swing their jaw from side to side to generate an efficient chewing action. And they too have diminished canines.

The small canine teeth of our earliest ancestors have been a talking point among palaeoanthropologists for many years, and they have been the source of many a fanciful hypothesis. One particularly popular idea was that because any self-respecting primate would not venture into open woodland, still less the savanna, unless he was adequately equipped with sharp rapier-like canines with which to ward off predators, *Ramapithecus* must have invented weapons to make up for his miserable canines. The idea fits in neatly with the suggestion that in order to exploit his new marginal environment the diminutive hominid needed tools (presumably sticks) with which to dig up roots. Tools very readily became weapons in the inventive mind of a hominid under threat of being dispatched by a hungry carnivorous cat. Or so the argument goes.

Ramapithecus may well have used the odd stick to prise food out of an inaccessible nook or cranny. And it may well have brandished branches either to impress a potential predator or, more likely, to impress each other, just as chimpanzees occasionally do today. But these activities are almost certainly unrelated to the animal's diminutive canines.

For illustration, let us contemplate the magnificent canines of male olive baboons. These sleek animals sport canines that may be up to four gleaming inches long in their upper jaws, and they can be razor sharp. This dental equipment is potentially lethal, and yet its owner will be one of the first to scramble to the safety of the trees at the approach of a predator. Only when there is absolutely no alternative does a male stand his ground and display a so-called threat yawn at an adversary.

Baboons' canines may well therefore be important on the odd occasion when an aggressive confrontation with a predator is unavoidable, but it is clear that their real function is to do with the males' principle evolutionary driving force: sexual success. This means that the animal that manages to fertilise more females than his fellows will produce more offspring to populate the next generation. And that, in terms of natural selection, is deemed success. Males are therefore in a constant, but usually subdued, state of competition with each other for the favours of the females. In the animal world this competition can take many forms: the peacock has its rainbow-coloured (and cumbersome) tail; a gorilla displays his superiority by his great body bulk and the sheen of silver-coloured fur on his back; a red deer stag carries bigger and more elaborate antlers on a bigger and better body; and the olive baboon has his gleaming canines.

This phenomenon, which Darwin recognised a hundred years ago and called sexual selection, leaves its mark on the animal world from end to end: it is the reason that the beautifully decorated birds are usually the males whereas the females, which are the prize that the competition is all about, are frequently drab; and it is the reason that males are often heftier than their sexual mates – they are out to impress the ladies and to persuade the competition that they are un-questionably superior. Sexual selection has obviously been at work in humans too: men, on average, are around 10 to 15 per cent bulkier than women; beards, which are present in some peoples but not in others, are also probably symbols of sexual competition, but ones that are usually removed in many western cultures; and men's canines, though not large, are definitely bigger than women's.

What of *Ramapithecus*? Having suggested that the change in its dental equipment is the result of ecological necessity rather than cultural invention, what can we say about its sexual selection? Nothing. At least, nothing that comes from direct evidence. But let's take just a little further the idea that our ancestors were making a living in pretty demanding environ-mental conditions, conditions in which, unlike in the forests, food was limited if not actually scarce.

First of all, *Ramapithecus* almost certainly was a social animal. This is not a wild guess because we know that most primates, particularly the higher primates, live in social groups (the monogamous gibbon is an exception, but there are good ecological reasons for that). The probable reason for living with your fellows is partly that it offers some protection against hungry predators, but mostly that it provides a ready-made education system for the young: primates have a lot to learn, and living in groups is the best way to achieve it.

There are, however, different ways of indulging in group living: the two extremes are, first, a mixed group of males, females, and young; and, second, the harem, with females, young, and one all-powerful male. The factors that mould animals into one way of life or another are many, but include influences such as sexual selection and response to the wealth, or otherwise, of essential resources.

Let's look at the gelada again, not because we proffer it as the perfect 'model' for early hominids, but because we have to ask questions somewhere, and the gelada is as good as any, and better than most. (The idea of looking for complete living 'models' of ancient hominids is fruitless and misleading, but with caution one can explore limited models for specific pieces of behaviour.) As we said, geladas eke out a living in the highlands of Ethiopia where food is not super-abundant. It so happens that these monkeys live in harems, with groups of

33

perhaps four females being kept in order by a male that is almost twice as bulky as the average female. In addition, the male is tremendously hairy, sporting a voluminous mane, all of which contributes to his impressive looks (impressive not only to human onlookers, but also to his mates and to any young bachelor with designs on the women).

Because of the way evolution and natural selection work, the shape of a social group is determined by what is best for the individuals in it, best in terms of sexual success, that is. When food is scarce in a particular type of habitat, the individual's strategy for maximum sexual success can sometimes encourage the formation of harems: the females in the group don't have to compete with lots of males for the limited food, and the harem owner doesn't continually have to compete for the sexual favours of the females; there are the occasional and important confrontations over proprietary rights of the harem, of course. The groups of more or less discontented bachelors skulk off elsewhere, just waiting for an opportunity to grab some females of their own to set up new harems. Where you have harems you see exaggerated signs of sexual selection, or sexual dimorphism as it is called: the hairiness and big body of the gelada males is one example; so too is the masculine muscularity in gorillas and the hanuman langurs that live in Southern Asia – and in humans too.

Did *Ramapithecus* live in harems? Were the males much bigger than the females? And did the males have a thick coat so as to make them look even bigger, just like the geladas? It is possible, but we simply don't know. Isn't all this just an empty exercise then, you might protest? Up to a point it is, but only if what you are after is a rigid answer to all of our questions about *Ramapithecus* right at this minute. But if we can contemplate the past in an open-minded way and try to learn from the rules of biology, rules that are still being worked out by ethologists, then in the end we are more likely to get closer to the truth of what our ancestors were up to 15, 10, three million years ago, or whatever period we may ask about. For instance, there will undoubtedly come a time when enough fossils are collected so that it will be possible to answer the question, were the males bigger than the females? An answer to that takes us much further than a simple comparison of the physique of the two sexes; it gives us a posthumous glimpse of the social life of the animals.

Socialising, as we said, is nature's education. Youngsters grow up and, with the protection and guidance of adults and with the innate imaginative creativity of play, they learn about the world around them: they learn what to eat, where and how to find food, and what dangers to avoid; perhaps most important of all, they learn about each other. And the more there is to learn, the longer the infants have to stay in nature's school. Throughout human evolution there undoubtedly has been a steady extension of childhood: as we became more and more complex animals depending to an ever-growing extent on our intelligence and wit rather than on inbuilt programmes in order to make a living and cope with and manipulate the nuances of social life, the more we depended on education.

Probably the most dramatic thing to have happened to *Ramapithecus* is that it learned how to walk upright. We don't know how it got around the place before it adopted this highly unusual method of locomotion: maybe it moved smoothly on all fours, much as olive baboons do today; we don't know, and indeed, we shouldn't try to pin *Ramapithecus* down too precisely to what we can see in contemporary primates because, in all probability, the ancient hominid did its own thing. But what we do know is that by three million years ago our ancestors walked around the Pliocene landscape in much the same way as modern humans walk around modern cities.

By standing and walking upright *Ramapithecus* would have freed its hand wonderfully. It would have been able to carry things, to learn to throw accurately, to manipulate small objects with undreamed-of precision, and perhaps even to invent a language of hand gestures. A whole new world is opened up when a primate learns to walk on two legs instead of four, and the evolutionary force propelling *Ramapithecus* into this unique position must have been pretty powerful because the new stance demands radical restructuring of the pelvis and leg anatomy. We are talking here of *habitual* upright walking, rather than *occasional* bipedalism, something that all apes are capable of, inelegant though it looks.

That it happened we know. That there are considerable advantages to be had once one has stood up is incontrovertible. But *why* it should happen in the first place is a mystery because most of the advantages are apparent only when upright walking is very well advanced.

The arguments about bipedalism, as it is called, can be

slotted into one of two main categories: cultural and ecological, and so far neither is overwhelmingly convincing.

Tools are at the focus of one cultural argument. Some people suggest that the need to fashion tools is sufficient to demand that ancient *Ramapithecus* should be able to stand up. *Ramapithecus* stood up, that we know, but this particular line of reasoning doesn't. First of all the earliest stone tools don't appear in the archaeological record until relatively late, around three million years, and these are extremely crude, certainly not the work of hands tuned to fine precision. In any case, our ancestors were fully accomplished bipedalists by this time, and probably had been for several million years. Unless one wishes to propose a culture centred on elaborately carved sticks (which, of course, disappear without trace), then this particular culture-based argument for bipedalism should probably be dropped.

What of ecology? One suggestion has been that as the forests shrank *Ramapithecus* set himself up as a hunter on the open savanna, and as he was such a diminutive chap he got on much better by standing on his hind legs to peer over the tall grass to spot the prey. This argument too is a non-starter. First of all there is no evidence that our ancestors adopted a voracious taste for meat at this time, indeed the teeth tell us quite the opposite. And secondly, it is reasonably certain that *Ramapithecus* developed his new way of walking while he was still making a living in the forest fringes, or at least in woodland.

One attractive notion is that walking on two legs allows the animal to cover a greater range more effectively. If *Ramapithecus* really did have to search high and wide for its food and to have the wit to know what type of vegetation was ready to eat where and when, then it probably would have to exploit an unusually large home range. Walking on two legs would help, if it were cheaper than going around on four. But it probably isn't. Bipedalism appears to consume more energy per mile than quadrupedalism. Yet another simple theory is frustrated.

As a side issue to this main problem it is intriguing to speculate on a possible link between the energetically expensive business of getting about on two legs and the origins of human 'nakedness'. Whether or not it was a principal cause of bipedalism, upright hominids almost certainly covered more ground as they developed a more ranging lifestyle. They probably carried things too, such as food and implements. All of this is a hot business under the tropical sun, and the problem of overheating would have been made worse by the extra energy having to be burnt to fuel this new and expensive mode of locomotion. An efficient way of keeping cool would have been essential, hence the development of countless sweat glands and the dramatic shortening of body hair, both of which are very un-ape-like.

Such a development would almost certainly have plunged our ancestors into a nakedness more biologically significant than the simple lack of long hair or clothes: like most modern monkeys and apes, *Ramapithecus* very probably had pink skin, a characteristic that would have made our ancestor supremely vulnerable to the searing rays of the sun. Evolution's answer would have been an all-over blush of dark pigment cells giving the skin a protective brown colour. That early biological blush would have been lost in some human populations as some of our ancestors colonised colder parts of the world where a dark skin blocks the formation of vitamin D.

How bright was *Ramapithecus*? With little more than fossil teeth and jaws for evidence it is not easy to say, of course. But there is one tiny chink of evidence we can point to in these meagre petrifications. In some of the fossil specimens one can detect that the grinding millstone teeth, the molars, are worn to different degrees: starting from the front of the jaw, the first molar is worn most, the second is ground down slightly less, and the third is even more intact. The same type of wear pattern happened in the mouths of *Sivapithecus* and *Gigantopithecus*, but not to the same marked degree. What does it mean?

As we all know from experience, and sometimes to our agonising cost, permanent molars erupt through our tender gums during early youthful years: the first to come are at the front of the jaw, the last at the back. Now, if you were a little *Ramapithecus* your teeth would begin their millstone role as soon as they were properly through your gum; the wear begins almost immediately. The fact that there is a noticeable difference in the amount of grinding down inflicted on the teeth means that there must be a significant lapse of time between the appearance of the first and second, and the second and third molars. And this means a prolonged childhood – we think.

If the teeth do not lie, we can say that even 10 or so million years ago our *Ramapithecus* ancestors were making a living that demanded greater wit than, say, *Sivapithecus* and *Gigantopithecus*.

Gigantopithecus is a little odd. A big animal, it almost certainly

37

lived mainly on the ground and, like *Ramapithecus*, its large, flattened teeth tell us that it ate a tough diet. This massive creature must have worked itself into an evolutionary rut, however, because in India and China at least it hung on with much the same lifestyle into the late Pleistocene, just half a million years ago.

Ramapithecus and *Sivapithecus* were obviously neighbours, however, because their petrified remains are often found together. From its large incisors and prominent canines, we can tell that *Sivapithecus* had a different lifestyle from its smaller contemporary: it may well have eaten a lot more leaves and soft fruit, in which case it may have spent a lot of time in the trees, casting occasional glances at its cousins going about their business on the forest floor below. Indeed *Sivapithecus* may have been the reason that *Ramapithecus* was forced to try his luck on the ground and outside the forests.

CHAPTER 3

The fossil void

When you stand in front of the fossil-hunting camp at Koobi Fora and look into the setting sun you face a magnificent range of mountains darkly silhouetted against the brittle blue sky. This is no ordinary range of mountains. This is part of the western wall of the Great Rift Valley.

Although the Valley widens by an apparently unimpressive one millimetre a year, in a few tens of million years from now the Valley will be no more: it will be a sea. The warm blue waters of what is now the Indian Ocean will begin creeping up the Valley and this unavoidable prospect of 'The Great Rift Sea' reminds us of the certain transience of our species' occupation on the planet earth. Sometime, sooner or later, *Homo sapiens* will become extinct.

How many people would be prepared to argue that their descendants will be surviving to bathe and fish in that new sea that 20 or so million years from now will carve its way into Africa? Probably not many, and the rules of biology support such pessimism. But no one will need reminding that in its gradual emergence from the animal kingdom humanity invented a new game, and that game is called culture. This is not the creations that hang on walls in art galleries or fill the pages of literary books. It is the stuff of human existence, the quality that allows us to impose our will on the world around us rather than having to respond to its every quirk and tremor. And it is the unique blend of biology and culture that makes the species *Homo sapiens* a truly unique kind of animal.

Every species of animal is, of course, unique in its way – in the way it looks and how it makes its living – for that's what the term species means. Humans are different, not so much for *what* we do – though it would be more than a little extraordinary to see, say, a baboon turning out printed circuits or driving a car – but rather for the fact that we can do more or less what we want. That is what having a highly developed culture really means. And it is the rules of this new game that will determine how we will survive as a species in the future.

The blend of biology and culture that is humanity is clearly an unusual quality in the animal kingdom because, in the terms of zoological classification invented by Carolus Linnaeus in the mid-eighteenth century, *Homo sapiens* stands alone as the only representative of the genus *Homo* and the family *Hominidae* (more commonly called hominids): we are deprived of any close relatives. We can claim the endearingly human chimpanzee as our closest cousin, but unlike us he boasts fellow members of his family, the *Pongidae* (pongids): they are the other great apes, the orang-utan and the gorilla. Something unusual happened in our past, something that, while making us the truly extra-ordinary animal that we are, at the same time left us bereft of living relatives; unlike in the past, no other hominid shares our planet with us now.

Why? What made us special? How did our ancestors break the bonds of biology to take on the cloak of culture? And what does it mean for our future? We ask these questions partly because one of the unusual qualities of the human mind is that it is intensely inquisitive. But we ask them too because in their answers lie the key to the long-term survival of *Homo sapiens*. To be sure about our future we must know about our past.

At one time it was fashionable to think of the emergence of modern man from an ape-like stock as having followed a simple ascent up a straight staircase, each step being marked by the appearance of more and more advanced features: the design was steadily improved, so to speak. Indeed, at one time it was possible to squeeze the fossil evidence into this kind of 'single species' framework, as it was called. The notion is now largely abandoned. All the evidence goes to show that the issue of the *Ramapithecus* which flourished 13 million years ago gave rise to the family of three hominid types (*Homo* and the two australopithecines) around five million years ago. But because, for several reasons, the period between eight and about four million years ago is a virtual fossil void, we can only guess what our ancestors were up to then. Nevertheless, the structure of human evolution is clear: we start off with an ancestor about 15 million years ago; by three million years ago the hominid stock has proliferated to produce a number of related creatures, one of which is the *Homo* line that eventually becomes modern man; and during the further evolution of the *Homo* stock the rest of the hominids die out.

The story of that initial proliferation followed by an evolutionary pruning is the story of our origins.

If we are to be swayed by the weight of current fossil evidence then it is quite possible that the most dramatic events in human evolutionary history were played out on the African continent. Modern humans colonise practically every corner of the globe, from tropical Africa to the Arctic circle, a distribution that speaks volumes for the adaptability that has made us such a successful creature. By contrast, our ancestral woodland ape appeared to prefer a warm climate: his remains have been unearthed in Africa, southern Asia, and parts of Europe that were much warmer eight to 15 million years ago than they are now. The hominids of between two and five million years ago, however, seem to have been much more choosy: so far they have turned up only in Africa, nowhere else.

This apparent evolutionary parochialism seems to be remarkable, and indeed it may be only a quirk of fossil hunting. It is true that there are not many good exposed fossil sites of the right age in Europe and Asia: hominid remains may in fact be buried deep and inaccessible in these continents; or perhaps they may have been swept away as erosion gnawed through three million year old deposits. Fossil discoveries outside Africa may come, and a number of people have hopes that some sites in Pakistan where specimens of *Ramapithecus* have been discovered will, in younger deposits, contain the later hominids. But so far it remains a fact that, in spite of considerable effort, no fossil hominid dating between two and five million years has been found anywhere but Africa.

Long before the first fragmentary fossil evidence came to light in Africa, Charles Darwin suggested that Africa was the cradle of mankind. In a book called *The Descent of Man*, a book he published in 1871, 12 years after his historic *Origin of Species*, he argued: 'In each great region of the world the living mammals are closely related to the extinct species of the same region. It is, therefore, probable that Africa was formerly inhabited by extinct apes closely allied to the gorilla and chimpanzee; and as these two species are now man's nearest allies, it is somewhat more probable that our early progenitors lived on the African continent than anywhere else.' Only time and more years of patient searching through the right deposits outside Africa will prove Darwin right or wrong in his prediction. But at the moment the signs are in his favour.

One guess about the putative African origin of the hominid family is that the diversity of environments generated by the geological unrest of the Rift Valley might have been important. This *might* have been one reason for the apparent absence of Pliocene hominids outside Africa, but a possibly more persuasive argument invokes climate.

During the past 50 or so million years there appears to have been a slow but steady fall in global temperature, dropping by perhaps as much as 20°F during that time in temperate regions. As well as this continuous gradual cooling, there were also rhythmic fluctuations: in other words, instead of drawing a *straight* line sloping down towards the present showing the steady cooling, the line should be *wavy*. Each time the wave rises, the tropical region expands; each time it comes down, the tropics shrink.

One expansion of the tropics may well have been around 15 million years ago, shortly after the African continent joined again with Eurasia, the time when *Ramapithecus* and other tropical mammals wandered into southern Eurasia. We do not know yet because no one has examined the fossil and geological evidence, but it could be that during the crucial period when *Ramapithecus* began to beget his hominid descendants, the tropical zone may have shrunk again, making southern Eurasia unsuitable for tropical animals such as the hominids. *Ramapithecus* may not have had the opportunity to exploit southern Eurasia, simply because it was no longer a tropical zone. Probably, however, no *single* factor was operating in making Africa the cradle of mankind: more likely is that there was a web of interacting influences which, for the moment, remain obscured from our view.

Each time someone unearths a fossilised fragment of one of our ancestors that had been buried deep in the dust-dry deposits laid down perhaps two million years ago in Africa, we move one small step closer to some of those answers. Each time an ancient campsite once again feels the heat of the tropical sun as it is carefully excavated by an archaeologist, we can hope to fill in one or two more tiny details about our history. Unfortunately, the path of human evolution is littered but sparsely with meagre clues about our ancestors: stone tools, fragments of a skull perhaps, pieces of a limb bone, part of a foot, half a jaw, occasionally an almost complete skull – and lots of teeth, scores of teeth! This is what we find when searching through the now-buried deposits where our ancestors thrived.

One thing that a newly emerging branch of science, known as taphonomy, has taught fossil hunters in recent years is that when an animal dies the chances that its bones will remain intact long enough to be buried and become fossilised are vanishingly small, and this applied to the corpses of ancestral humans just as much as to other animals. In fact, fossil hunters don't need to be told this as a scientific fact, they know it to their cost and frustration. It is a chastening fact that if someone went to the trouble of collecting together into one room all the fossil remains so far discovered of our ancestors (and their biological relatives) who lived, say, between five and one million years ago he would need only a couple of large trestle tables on which to spread them out. And if that were not bad enough, a not unusually commodious shoebox would be more than sufficient to accommodate the hominid fossil finds of between 15 and six million years ago!

Nevertheless with a confidence that may strike the un-initiated as something close to supernatural – if not to plain madness – prehistorians can now construct a view of human origins that is anything but crude, and may even bear some resemblance to the truth. For although the fossil evidence is so pitifully inadequate what we have now seems like untold riches. Let us look, therefore, at the meagre contents of a reluctant fossil record.

At a site just a few miles west of Olduvai Gorge, Mary Leakey and her colleagues found fragments of jawbones from individuals who may have been our direct ancestors and who were living almost four million years ago; and she is even now excavating the fossilised footprints that (probably) these creatures left in soft volcanic ash, giving us, literally, footprints in the sands of time; a joint French/American expedition, working in southern Ethiopia at the point where the great Omo river heaves its orange silt into the jade green waters of Lake Turkana, have uncovered tiny fragments of quartz that, almost unbelievably, turn out to have been tools made and used by hominids living around two million years ago; further north, in Ethiopia, Don Johanson and his colleagues discovered the remarkably complete skeleton of a three milion-year-old little lady known as Lucy who belonged to a frankly unusual species of hominid; Johansen and his team have also uncovered the remains of a group of some 30 or so individuals whose bones became buried in a lakeside gully almost three and a half million years ago; from a spot close to where Lucy was found archaeologist Helene Roche has discovered stone tools which were struck around two and a half million years ago, making them the oldest undisputed stone implements in the archaeological record. And from the eastern shores of Lake Turkana there is the famous – some say infamous! – skull known numerically as 1470, the most complete skull of a human ancestor who lived slightly more than two million years ago; and recently the lake shore deposits have been persuaded to part with a number of remarkably advanced skulls of individuals who lived there around one and a half million years ago, discoveries that have an important impact on our view of a particularly critical period of our history, the time when some of our ancestors left their native Africa and began to colonise the rest of the world.

These are just some of the discoveries that are helping transform our understanding of human pre-history. What the fossils tell us directly, of course, is what our ancestors and their close relatives looked like. Or rather, to be more accurate, they give us some clues about the physical appearance of early hominids which followed *Ramapithecus*, because until someone is lucky enough to come across a complete skeleton of one of our ancestors, much of what we can say about them is pure inference, guesswork.

Not only is there no single complete skeleton we can learn from, but we do not even have a big enough range of fossil fragments from which we could make up a 100 per cent identikit ancestor! If we were to try to piece together into a kind of composite skeleton the fragments we have of our direct *Homo* ancestor of about two million years ago the task wouldn't take very long, simply because there are so few pieces to slot into place: the product would be pitifully incomplete – a skull, possibly part of an arm, a couple of leg bones, perhaps half a foot, and little more.

Our reconstructed skeleton would not be much help to the average medical student needing to learn anatomy. But for pre-historians, surprisingly, it is enough. For instance, we can see that by this time in our ancestors' history the brain was expanding, the jaw was unique and quite unlike any ape's, and these creatures walked upright around the countryside very much as we do today. Yes, we can construct a useful image of what these individuals *looked like*. But what we are really interested in is *what they did*, how they made a living, what was important in their social life.

If bones only rarely become buried and fossilised, behaviour never does, and unfortunately, it is the behaviour of our ancestors that we need to know about in order to understand ourselves better. A few products of behaviour do remain with us of course: stone tools, shadowy features of ancient living sites, bones of animals that may have been included on a pre-historic menu, a fragment of sharpened ochre perhaps, and, quite late in the fossil record, signs of an awareness of mortality when our ancestors started to bury their dead – the list is long, but it does scant justice to a complex social organisation which, though not fully human, was certainly not simply animal. The behaviour of our ancestors was neither a primitive form of our own, nor was it just that of a sophisticated ape: it was a special adaptation of its own, and it is this that we are trying to construct.

If we were to go back to our trestle tables carrying the fossil remains and were to select those of hominids that lived between two and three million years ago, we could sort them into a number of categories.

The bones in the separate groups have a number of striking resemblances to each other, of course: after all, they belong to the same family, the hominids. Nevertheless, each separate pile represents a distinctly different type of creature. Now, because the fossil evidence is so sparse there is lots of room for inter-pretation, different interpretation. The problem is made more acute because in any population of animals there is a degree of natural variation in appearance – look around at your fellow *Homo sapiens* if you need a graphic example.

If this type of variation was large in extinct populations the differences in the bones they leave behind them may be so great as to fool scientists into believing that there had been several different species where in fact only one existed. And if that were not bad enough, it is amazing but true that palaeo-anthropologists still cannot agree on solid definitions of what characteristics qualify any particular bone for acceptance into one species of fossil hominid or another. So it should not be surprising that if we were to invite six researchers to sort out the fossils into what they considered to be appropriate piles, the chances are that each person's selection would be different. Certainly, some people would disagree about which pile particular fossil fragments should be assigned to. But more than that, there may well be arguments about how many piles should be created in the first place. In other words there would be differences of opinion about what type of creatures made up the hominid family between two and three million years ago, and about what type of hominid some of the fossil remains belonged to.

The essential point, however, is that at this time more than one hominid thrived on our earth. As we said, some people will not agree, but we suggest that in East Africa at least three different hominids lived more or less side by side: first of all there is our ancestor, whom we call *Homo habilis* (which, roughly translated, means 'able man'), who stood upright, was between four and five feet tall, and had a notably large cranium; next is *Australopithecus boisei* (robust southern ape), who was about the same height as *Homo habilis*, but was much less athletic looking and had powerful jaws in a thick-set head; closely related to *Australopithecus boisei* is a smaller individual, near to four feet tall, whom we call *Australopithecus africanus* (African southern ape); there may also be a fourth and very shadowy group which is distinguished by the presence of Don Johanson's Lucy, and it may well be a remnant population of our ancestral woodland ape-like creature, *Ramapithecus*. (Johanson has named Lucy *Australopithecus afarensis*, and he adds into this species his family group and Mary Leakey's recent discoveries near to Olduvai. Johanson sees *afarensis* as the ancestor of both the *Homo* and *Australopithecus* lineages.)

In spite of the many similarities between the family members, they made their livings in different ways, a division of resource exploitation that was necessary if they were to coexist. (We are not suggesting that at some time these creatures sat around and decided how best to divide up the ecological opportunities; their different ways of subsisting were a matter of evolutionary divergence.) This state of coexistence must have continued for a very long time, but something happened to change it all: by about one million years ago only one type of hominid remained, a hominid that eventually gave rise to *Homo sapiens*. The other three slipped into extinction; they left no descendants. We are the sole surviving member of the hominid family.

46

We therefore have an image of our Miocene ancestors scampering on all fours (or however they moved) into the fossil void eight million years ago, to emerge five million years later, head erect and striding upright. And whereas just one ancestor went into the gloom, four progeny come out: *Homo habilis*, *Australopithecus africanus*, *Australopithecus boisei*, and a remnant *Ramapithecus* of some sort.

Four hominids

One of the first impressions made by the Lake Turkana country, round the camp at Koobi Fora, is of an arid wasteland, surely incapable of supporting all but the most meagre forms of life. Indeed, in the dry season the unrelenting sun bakes the brown sandstone to scorching temperatures, the hot air above dancing to a frenzied tropical tune. Life literally seems impossible. In fact, the lakeshore teems with activity: more than 300 species of bird live there, including flamingos, pelicans, spoonbills, stalks, and geese; crocodiles and hippos abound; on the grassy shores herds of zebra, grants gazelle, and topi are as numerous as you'll see anywhere in Africa; ribbons of trees that follow seasonal river courses shelter rhinos and pigs (one member of the research team was so impressed by the interest a couple of pigs showed in his geological survey that he named the area 'wart hog canyon'); meat on the hoof is certainly plentiful enough for numerous lions, wild dogs, hyaenas to make a very satisfactory living. And less than 100 years ago, elephants and water buffalo were common at the lake too.

The area today is a patchwork of barren sandstone terrain, grassy shore margin, bushland, and ribbon forests skewering their way to the lake margin alongside numerous river beds. The only standing water, apart from the alkaline lake, is a spring at the foot of the eastern highlands at a place called Derati. Once every three weeks workers from the Koobi Fora camp drive a lorry to Derati and collect more than 800 gallons of the crystal clear water. (Unfortunately the virgin freshness of the spring water vanishes soon after it gets back to the camp as it is stored in disused fuel drums!)

Yes, the lake shore is semi-arid, but it is a place rich in animal life, and, what is more, following the brief but often torrential rains in April, May, and November, the barren landscape is transformed with the sudden blossoming of flowers. Within a few days of the rain a carpet of small purple flowers hugs close to the sandy shore, and where the sand gives way to soil away from the shore the carpet turns from purple to yellow, a profusion of tiny buttercup-like flowers that follow the sun as it arcs from east to west. Landing at the Koobi Fora airstrip during the post-rains flower gala is a magic experience. Anticipating the rains by two to three weeks the Acacia bush, known locally as wait abit, comes into bloom. For most of the year the bushes stab the air with row upon row of as vicious thorns as you're ever likely to encounter, but in the period before the rains the bushes undergo a plant's equivalent of a personality change and they drape themselves with a beautiful soft covering of white flowers: they look like bushes after a snow storm.

One lakeside plant that is spectacular less for its flowers than for its medicinal properties is *Sansaveria*, a succulent that has the local name of Olduvai. It grows in many parts of the Rift Valley and it is from the plant that Olduvai Gorge gained its name. Twisting the thick, ridged leaves produces a clear liquid which has remarkable powers for healing open wounds, acting both as an antiseptic and as a natural bandage in binding the wound together. Nomadic peoples in the Rift use Olduvai, as too have two generations of Leakeys following accidents during field excursions. It is certainly preferable to anything that twentieth-century pharmaceuticals can offer. Bill Montagne, director of the Oregon Primate Center, benefited from the plant's powers after a bad fall near to Koobi Fora in 1976, and so impressed was he with the results that he is currently having the Olduvai's liquid scientifically analysed to discover the nature of the active agents. One wonders whether our ancestral hominids had discovered the properties of the plant, for it undoubtedly was growing there two million years ago, because in 1972 Richard Leakey's wife, Meave, helped to produce evidence that *Homo habilis* was living near the lake at this time.

1972 is a year that will always remain special in her memory. For a start it was the year in which she gave birth to her first child, Louise. And second, she was principally responsible for achieving the daunting task of reconstructing a tremendously important hominid skull that had been discovered in deposits close to two million years old at Koobi Fora. As things turned out the two events – Louise's birth and Meave's role in reconstructing the skull – were not entirely unconnected.

Louise was born in a Nairobi hospital in March. Four months later she became the youngest member of the Koobi Fora research project when, safely cradled in Meave's arms, she touched down in a single-engined plane at the camp's airstrip. Meave, who has a doctorate in primate anatomy, had been doing important work on the fossil animals at Koobi Fora – particularly on the monkeys – and she was determined to continue in her new state of motherhood. Louise still makes regular visits to Lake Turkana, and she now has a younger sister, Samira, as a playmate.

Four days before Meave and Louise flew into Koobi Fora (with Richard) in the late summer of 1972, Bernard Ngeneo had been out on a routine survey about 13 miles northeast of the camp. It was an area where the brief but fierce seasonal rains cut steep gullies in the ancient sediments, very typical of much of the terrain around that part of the lakeside. One particular gully had been surveyed many times and there was now a well-worn path through it. Close to the path lay a few fragments of bone, mostly very small. Many people must have seen them before, but no one took the trouble to stop and examine them closely. That day Ngeneo did stop, for no particularly special reason. Instinct of a fossil hunter perhaps? In any case, he looked at them carefully and noticed that one of them appeared to be part of a cranium, a hominid cranium.

The piece of bone that had caught Ngeneo's attention so strongly was part of the front of the cranium, and it was clear from the curvature of the fragment that it had once been part of an unusually large brain-case. Wind and rain had started their crude excavation of the skull at least a year before Ngeneo noticed it, and the fragile fossil had begun to disintegrate. However, judging by the collection of exposed fragments as they lay on the surface, there seemed to be an excellent chance that more pieces would be found and that they could be reconstructed so as to reveal whether or not Ngeneo had indeed found a truly unusual hominid.

There was a keen sense of anticipation at Koobi Fora, and the excitement grew as more and more fossil fragments were gently sifted from the soft sand and taken back to the camp. Meave soon took over the task of putting the pieces together: 'I had to stay in the camp anyway, to look after Louise,' she says. Eventually there were more than 300 fragments discovered, some as big as a matchbox, but most the size of a finger nail or smaller. 'It's like doing a three-dimensional jigsaw in which you have no idea of the size, the shape, how many pieces there should be, or indeed if any of the pieces are missing – it's very exciting.' For most people, doing a jigsaw of this nature would be a nightmare. For Meave it was a tremendous experience: 'You feel exhilarated when you get a piece to fit. You think about it all the time. You dream about it. Suddenly in the middle of the night you get an idea of where a particular piece might fit, and you have to dash off and try it.'

For five weeks Meave stuck to her task, with the occasional more or less inspired help from others in the camp. For those five weeks she kept in her head countless details of the size and shape of pieces, the internal pattern of the fractured fossil, mentally shuffling them to try to slot a recalcitrant piece into place. In the welcome cool of the thatched banda that stands just 40 yards from the gently lapping green waters of the Lake she gradually pieced together the impossible jigsaw, occasionally taking time off to feed Louise or simply to rest: 'You can't keep at it all the time, especially if it isn't working out; some days you don't get a single piece to fit, and that *can* be frustrating.' Meave has more than patience and persistence to help her. She has a real gift for jigsaws: 'I always liked jigsaws, and as a child I used to turn them upside down if I found them too easy.'

Very soon after the cranium began to take shape it became clear that the earlier suspicions that it was something special were correct: its brain was large (close to 800 ccs) and the face was smaller and less protruding than its more primitive australopithecine contemporaries. For the final stages of the reconstruction the cranium was taken back to the museum in Nairobi where British anatomist Alan Walker worked on it. While the cranium was in these final stages of reconstruction the skull was examined by Louis Leakey in his room at the museum. He was delighted because it helped vindicate his belief that the genus *Homo* was born at least two million years ago, much longer than most people were prepared to accept at the time. Soon after he saw the skull Louis left Kenya for a lecture tour of Europe and the USA. He died of a coronary in London a week later, on 1 October.

For some reason the cranium never acquired a nickname. During those weeks of August and September at Koobi Fora it was always referred to as 'Ngeneo's cranium'. Eventually it came to be known prosaically by its catalogue number at the national museum: 1470. It is surely of the genus *Homo*, a member of our direct ancestors who lived and died close to the ancient lake shore around two million years ago.

1470 is important not just because of its age, but because, after its careful reconstruction, it is the most complete specimen of its type: its cranium and face are virtually intact, but its lower jaw (the mandible) is missing. Perhaps a hungry hyaena or other scavenger tore the lower jaw from the rest of the head shortly after 1470 died. We shall never know. In any case, in spite of an extensive search through the sediments, the jaw was nowhere to be found. We know that shortly after 1470 died a mouse nibbled its skull, leaving tiny teeth marks we can see today.

We can now be fairly confident that the reconstructed 1470 looks very much as it did when, close to two million years ago, it lay gleaming white in the tropical sun before being buried at the beginning of its journey into the fossil record. And this is because there is a rule about reconstructions at East Turkana: only pieces that unequivocally fit together shall be used. Although the temptation is great, there is very little value in creating a reconstruction that is in truth just pieces of fossil bone floating in a sea of plaster of paris: not only is this not helpful, it can also be dangerously misleading.

The fossil record is certainly not profligate in giving up its treasures to us, but we must avoid misusung the offerings we do receive. Our rule means that, more than five years after 1470 turned up, there is a drawer in the National Museum's 'hominid room' that still contains some of the ancient jigsaw: the pieces are taken out from time to time, tested against the gaps in the reconstruction, and put away again; there they will stay until they fit with certainty somewhere in the skull.

We know that when 1470 and his fellows were alive the Turkana area was wetter than it is now: the lake was higher, with a shore perhaps 10 to 15 miles east of its present position; and the vegetation was more lush, offering an exaggeration of the current patchwork, with rolling grasslands and gallery forests. 1470 would undoubtedly have seen many of the ancestors of today's plants, birds, and animals, and many more besides.

This type of mixed waterside ecology is typical of where we find hominid fossils in East Africa: the Hadar, lower Omo, and Olduvai all fit into this pattern. Most animals need to be near enough to water to be able to slake their thirst at least once a day. Compared with most animals, humans, and probably the early hominids too, are particularly dependent on water. Choosing a lake margin or river side as a place in which to make a living would therefore have been sensible for our early ancestors. The archaeological record is bound to be biased, however, because in places where there are no gently lapping lakeside waters or silt-laden streams or rivers, there is virtually no chance of hominid skeletons being fossilised, no chance of our finding traces of them eons later.

Although the creatures that 1470 lived among were distinctly prehistoric, many were very similar to today's, and for most at least we can see modern 'models'. Except for two: *Australopithecus africanus* and *Australopithecus boisei*.

These two hominids both walked upright: apart from ourselves, there is no modern animal that habitually strides around on two legs leaving the front limbs free.

Therefore, in the days of 1470 some two million years ago, three different types of hominid were an important part of the lakeshore livestock, with possibly a fourth (remnant *Ramapithecus*) being rather less prominent.* They probably experienced eyeball to eyeball contact with each other. Probably they shared the same water sources, sometimes digging down to fresh water in the dry bed of a seasonal stream just as the Dassanetch people do today, sometimes trekking to Derati, or perhaps going to another spring that no longer exists. Perhaps they used neighbouring trees for shelter from the blistering noonday sun. Or perhaps they assiduously steered clear of one another, one type inhabiting the gallery forest, another living mainly on the grasslands, while the third stuck close to the lake

* One problem that has beset palaeoanthropology for a very long time, and will probably continue to do so, is the naming game. There are often differences of opinion as to whether a particular fossil should be given the status of the genus *Homo* or whether instead it is better described as being in the genus *Australopithecus*. And when we talk of *Australopithecus robustus/boisei* (or *Paranthropus*) on the one hand, and *Australopithecus africanus/gracilis* on the other, we are referring to species that are thought to be similar enough to be classified under the one genus. The names should not be thought of as firm as yet, and an increasing number of scientists are favouring the classification of the australopithecines into two genuses – *Paranthropus* for the robust species, and *Australopithecus* for the gracile individuals.

shore, with the fourth skittering in between. We simply do not know, but for sound biological reasons it is fair to guess that the truth lies closer to partial integration than to strict segregation.

The fossil record at Lake Turkana suggests to us that the three principal hominids coexisted for at least one million years and probably twice that long, the australopithecines slipping into extinction around one million years ago. For at least the early part of their tenancy on the shores of the Jade Sea the two australopithecines and *Homo* were physically rather similar: it was their lifestyles that separated them. Animals can coexist for long periods of time only if they exploit significantly different resources: chimps and gorillas get along together because, though both are basically plant-eaters, gorillas concentrate on leaves and stalks whereas chimps like to include fruit and other diverse items in their menu. It is a fair guess that, though they overlapped to some degree, the food tastes of the three hominids were significantly different. They had to be, otherwise natural selection would have weeded one or other of them out much sooner.

As time passed the behaviour and social organisation of our *Homo* ancestor became more and more distinct from their hominid cousins: they were gradually leaving behind a primate subsistence *ecology* and replacing it with a pre-human food *economy*. We shall talk more about this in a later chapter.

So, what did our lakeside hominids look like? Let us start with *Australopithecus boisei*, the creature that provided the first major fossil prize in the lakeside deposits. This was in the summer of 1969, the first year of intensive exploration at the fossil-rich site. During a reconnaissance trip the previous year three members of the team – Kamoya Kimeu (who is deputy leader of the project), Nzube Mutwiwa, and Mwongela Muoka – each found a fragment of an australopithecine jaw. These tantalising discoveries, plus the obvious fossil potential of the vast sandstone deposits (animal fossils littered the surface just waiting to be picked up) confirmed the suspicion that this would be a rewarding place to look for signs of our Pliocene and Pleistocene ancestors and their relatives. And so it was.

It was during a long camel-back trek inland that the first *boisei* cranium was found in 1969. Three days out from the cool lakeside camp at Koobi Fora the camel party rested for an overnight stop before setting out finally for the border with

Ethiopia which could be reached the next day. But that objective was never achieved because the following morning the *boisei* cranium turned up.

After a 6 am breakfast of a large cup of sweet tea, an on-foot survey of the area began. By 10 am a pretty thorough survey had been achieved and thirst was driving the prospectors back to the camels by the shortest possible route. On the way the cranium was spotted lying upright in the dry bed of a sand river. Thoughts of thirst now completely gone, the virtually intact gleaming white cranium was as arresting as any vision. Every fossil hunter dreams of such a discovery. And there it was, a dream come true. Another few months and the dream would have been shattered as the seasonal rains crashed down the channel.

Running down the midline of the cranium was a crest of bone (called the saggital crest) which acts as an anchor point for the muscles that power the massive jaw (which, inevitably, was missing). By scooping away some of the sand around its base the extent of its wide cheek bones (also muscle anchors) became visible, emphasising the huge, flat, protruding face, a feature that contrasts with its small brain (just over 500 ccs). Its eyeless eye sockets were crammed with sandstone matrix, giving the skull a ghostly blank appearance. But around two million years ago this creature would have looked out on the varied countryside running down to a lake vastly bigger than today's. This was *Australopithecus boisei*, and it was the first one to be found in Kenya.

We know from subsequent discoveries of *boisei*, that its teeth and lower jaw were of impressive proportions. Set in a massive mandible and behind a row of tiny incisors and small non-protruding canines were enormous millstone-like molars. The premolars too were unusually large and molar-like, presumably to aid in whatever grinding process they were designed for. Some of the molars are up to one inch long from front to back, and in most jaws that are found they are worn flat with years of use.

The discovery of *Australopithecus boisei* at Lake Turkana came as one of those uncanny family and historical coincidences, for it was almost exactly 10 years to the day since Mary Leakey found the first *boisei* cranium at Olduvai Gorge. (By another coincidence, Mary just happened to be visiting the Koobi Fora camp at the time of the discovery!) The differences were, however, that during its passage into the fossil record the Olduvai cranium had shattered into more than 400 pieces, and that Louis and Mary Leakey's searches for hominids had been going on since 1931. During almost 30 years they had found wonderful examples of ancient stone tools in the deposits of a now extinct lake, but no trace of the hominids who had lived there. Mary's discovery in August 1959 broke the spell, and it was followed a year later with the find of a partial cranium (again shattered) for which the name *Homo habilis* was later invented.

By 1959 several examples of another type of robust australopithecine had already been excavated from the Swartkrans and Kromdraai caves in the Sterkfontein Valley of South Africa. Ever since 1924, when Professor Raymond Dart discovered a fossilised infant hominid that was christened the Taung child, excavations have been going on at a number of cave sites in South Africa. Although many prehistorians were unimpressed by the Taung child (they were still besotted by the infamous Piltdown skull, which in 1955 was exposed as a fraud constructed from a relatively modern human skull and an orangutan's jaw), Dart recognised his discovery as distinctly hominid-like. He called it *Australopithecus africanus*, which means African southern ape.

Dart was joined in his searches by a retired doctor called Robert Broom; later John Robinson and Phillip Tobias also joined in. There are now five major cave sites, the other three being the Sterkfontein caves which are almost within sight of Swartkrans and Kromdraai, the Makapansgat caves to the northeast, and the Taung caves (the resting place of the Taung baby) in the southwest. The cave deposits at Sterkfontein, Makapansgat, and Taung contain only *Australopithecus africanus*, and the australopithecine discoveries at Swartkrans and Kromdaraai have been of the robust version. So far the australopithecine cousins appear to have been unwilling for their bones to finish up in the same resting place.

A South African scientist, Elizabeth Vrba, has suggested that perhaps the segregation is the result of differences in local ecology: she looked at the different types of fossil antelope bones that accompany the hominids in their cave deposits, and she noticed that the animals that are known to have preferred open grassland (such as springbok and gazelles) are most often with the robust australopithecine, while its gracile cousin

56

mostly finished up in caves with animals thought to prefer more bush cover.

Vrba's suggestion is so far just a tentative hypothesis, and it will probably stay that way for quite some time because, as with the South African cave sites as a whole, there are many uncertainties. The biggest one of these is that at the moment there is no certain way to be able to say how old any particular fossil is.

The hominids may occasionally have lived in the caves, littering the place with bones of animals they brought in for supper, and leaving their own bones when one of their members died. But for the most part it seems certain that hominid bones found their way into the caves either because they were swept in by streams, or more likely that they were the supper of leopards, hyaenas, sabre-toothed cats, and lions, all of which lived in and around the caves.

In a brilliant analysis of the Swartkrans cave system, Bob Brain has shown how it is likely to be a refuse dump for the left-overs of leopards' suppers. From the collection of antelope bones in the caves, it seems that when the system was open the surrounding countryside was rolling grassland; there would have been very few large trees to be seen, except for occasional wild fig or stinkwood trees which survived by growing in sheltered hollows, places that often must have concealed the vertical entrance shafts to underground caves. This is very much how the Sterkfontein Valley looks today. We know that leopards lived in the valley at the time, and we know too that, when they make a kill, leopards take the precaution of storing their prize in large trees out of reach of over-attentive hyaenas and other scavengers.

Brain suggests that in times past when australopithecines occupied the valley they too would occasionally have finished up hanging lifeless in a leopard's arboreal store-cupboard waiting to be consumed. And as with other meaty items on the carnivore's menu, bits and pieces would have dropped off from time to time, some of which would have rolled into the cave shaft to collect with the other bones and rocks below. Ten bones a year is all that was needed to produce today's accumulation. The scenario is convincing.

Over the ages the composite of wind-blown dust and animal bones hardened to form solid limestone rock. And it is from this that the precious fossils now have to be hacked. A number of people have attempted to put an age on these rocky composites by indirect dating methods. Although there can at the moment be no certainty on these figures, it seems that Sterkfontein and Makapansgat were open between two and three and a half million years ago, with Makapansgat slightly older (around three million); Kromdraai seems to span the period between one and two million years ago; Swartkrans may be somewhere between one and a half and just over two million years old; and Taung, the site that started it all, may cover somewhat discontinuously the time between just less than one million to about three million years ago.

These uncertainties are bad enough, but as Bob Brain shows, they may be further complicated by natural excavation and subsequent refilling. For instance, at the Swartkrans caves the 'old' deposits were partially dissolved by rainwater pouring in through the shaft; this network of channels then filled up again with younger bones and rocks. The end-product is a perplexing intermeshed lacework of young and old, transforming an already difficult dating problem into a chronometric nightmare.

One factor that has made the East African sites particularly important is that the fossils discovered there can be dated reliably, and the means by which this is done is the ash that periodically tumbled out of the Rift Valley's many volcanoes. All the fossil-rich deposits from Olduvai in the south to the Hadar in the north are interlayered with volcanic ash, just like many layers of jam in a multiple sponge cake. We can determine when the ash spewed out of the volcanoes by a number of sophisticated geophysical techniques.

So, if we find a fossil entombed in deposits *underneath* a layer of ash that is two million years old but *above* another layer that was laid down two and a quarter million years ago, then we know that the bone was buried between these two dates. In practice we can usually be much more specific than this because the extremely active volcanoes belched out their gas and ash more frequently than every quarter of a million years. It is a poignant thought that during many a Pliocene evening the hominids on the shore of Lake Turkana would have gazed with wonderment at the staggeringly beautiful sunsets enhanced by gas and dust from volcanic eruptions, eruptions that in the future would help us reconstruct their family tree!

As well as helping to point up the importance of accurate dating in fossil hunting, our brief excursion into the South

African caves shows us an interesting phenomenon concerning the robust australopithecine. In South Africa, where they were first discovered, these creatures were certainly hefty fellows, standing close to five feet tall, with a generous endowment of weighty muscle. Eventually he was named, aptly, *Australopithecus robustus*.

As it turns out, all the robust australopithecines that lived in East Africa were built on the same pattern as their South African cousins, but they were significantly bigger – they were hyper-robust. Indeed, when Louis Leakey saw the fragments of the cranium that Mary had discovered in 1959 he decided that it was sufficiently different from the South African animals to merit not only a new species name, but a new genus too: he called it *Zinjanthropus boisei* (East African man). Zinj, as the skull is often called, also acquired the nickname of nut-cracker man, because of his massive teeth.

Why the East African robust australopithecines were so much bigger than those in the south we simply do not know. Evolutionary pressures often do encourage species to become bigger if ecological circumstances allow, so perhaps conditions were more to this robust creature's likings in East Africa. But it is only a guess. In any case, the animals are very similar, and in recognition of this the East African animals are now called *Australopithecus boisei*, that is the same genus as their South African cousins, but a different species. If they really were different species rather than, say, sub-species (and time and bitter experience is teaching us that it is unwise to be dogmatic about such things) then a useful contemporary example is the difference between the familiar chimpanzee (*Pan troglodytes*) which lives in West and Central Africa, and its smaller cousin the rare pygmy chimp (*Pan paniscus*) that is barely clinging on in Zaire. These two types of chimp *are* distinctly separate species: they don't successfully interbreed. *Boisei* and *robustus* *may* have been genetically distinct species, but at the moment there is no way for us to rule out the possibility that they were simply geographical variations of the same species.

As in South Africa, the robust australopithecines at Lake Turkana were accompanied by the more gracile version, *Australopithecus africanus*. But, unlike the southern sites, we know that creatures were at the lakeside in the same places and at the same time. (The more varied habitats at the lake may have offered *some* degree of ecological separation.) *Australopithecus africanus* stood a good foot shorter than his more bulky cousin, and he lacked the bony anchorage for jaw muscles on the top of his skull. His brain on average was 450 ccs, almost 100 ccs smaller than *boisei*'s, but this does not mean that he was dimmer; he just had a smaller body to control. Like *boisei*, his face was large and flat, but the teeth were somewhat different: in spite of being a much smaller animal, his incisors and canines were about the same size as *boisei*'s, but the molars were much smaller. Nevertheless, the molars were relatively large for a creature of its modest size.

Were the australopithecines hairy? Was *Homo habilis* slightly less hairy, just to give it a hint of human respectability? Certainly, all the portraits ever painted of our ancestors show this kind of pattern. But as no artist has ever seen a living early hominid, and as we haven't any way of knowing whether they were naked or not, it will remain a favourite topic of after-dinner speculation and fantasy for evermore.

Judging by the state of evolutionary divergence of the hominids around three million years ago, we can guess that descendants of *Ramapithecus* began taking their separate evolutionary pathways some time around five or six million years before the present. Perhaps the initial products were a proto-*Homo* and a proto-*Australopithecus*, with *Australopithecus* dividing again later to give the *africanus* and *boisei/robustus*. We have no way of knowing as yet, and it may prove impossible to determine because the closer one moves towards that initial divergence in terms of the fossils you find, the more difficult it is to distinguish one hominid from another: they will be all very similar to each other, and very similar to their ancestor too.

Assuming *Ramapithecus* in Africa around six million years ago found itself in an environment rich in ecological opportunities, we might reasonably expect that during its speciation the basic population would vanish, to be replaced totally by its descendants. Until a few years ago this certainly seemed to be the case. But a number of curiously primitive jaw fragments kept turning up at Koobi Fora, and they posed a problem: it simply wasn't possible to slot them neatly into the accepted structure of the time. Were they tantalising signs of a remnant *Ramapithecus* population? Possibly, but one could not be certain.

Then, late in November 1974 at the Hadar, Don Johanson made one of the most remarkable discoveries of the decade: it was nearing lunch-time on a scorching hot Sunday and Johanson had driven with one of his graduate students, Tom Gray, to try to relocate what he recalled from the previous year as a potentially interesting fossil site. Tired after an exhausting survey, they were taking a short cut back to their land-rover through the ancient lakeside deposits, which time and the elements had carved into countless gullies, when Johanson happened to glance over his shoulder and saw a fragment of a small arm bone lying on the ground. Johanson rapidly decided it was a hominid bone. Gray wasn't so sure: 'It was so small; it looked more like a monkey.' The two of them kneeled down to take a closer look. Gray suggested that they take it back to the camp to compare it with other bones. Suddenly he spotted three pieces of a hominid skull. 'We realised we might be onto something,' he recalls. The two of them stood back from the slope and immediately they started to see hominid bone everywhere: 'It was as if they were materialising right before our eyes,' Johanson says.

Johanson had stumbled on a skeleton that was about 40 per cent complete, something that is unheard of in human pre-history further back than about 100,000 years: Johanson's hominid had died at least three million years ago.

As luck would have it, all that remained of the cranium were a couple of small fragments, so we cannot say how big its brain was. But the lower jaw was intact; and there were ribs, arm bones, part of the pelvis, vertebrae, a thigh bone, and fragments of shin bones. For the first time it was possible to compare proportions of different parts of the body in a single individual. This was very important. For one thing, the arms are unusually long in relation to the legs in this individual, a clue that suggests that whatever else it did, the ancient hominid was probably adept at climbing trees.

Equally important though was the nature of the hominid. It was distinctly 'advanced' in that it had clearly walked upright. But the jaw had some persuasively primitive features, somewhat reminiscent of *Ramapithecus*. The jaw is distinctly V-shaped; the relatively large molars are flat; and the first premolar has a single cusp, a very primitive ape-like feature.

The shape of the pelvis suggests she was female, and the overall proportions make her very diminutive indeed, not much more than three feet tall. But she was not a young girl, she was fully grown (the teeth tell us that). One evening in the Hadar

camp shortly after the discovery, the conversation turned –
again – to the skeleton. Someone had put a tape on for music:
it was the Beatles' 'Lucy in the sky with diamonds'. Inspired
by the Liverpudlian music-makers, someone said 'Let's call
her Lucy'. And this is how the skeleton, whose catalogue
number is AL 288, became known to the world.

Having christened AL 288 Lucy, the next problem was to
decide what she is. She certainly doesn't fit into the 'obvious'
Australopithecus africanus category; her curious mixture of primitive and advanced features prevent that. Some people have
suggested that she is a separate species of *Australopithecus*. Indeed,
Johanson has suggested the name *Australopithecus afarensis*, the
predecessor, he claims, of all later hominids. But it is at least
arguable that Lucy represents a remnant population of
Ramapithecus that managed to survive the competition of her
evolutionary progeny. If this is so, then Lucy gives us a welcome
glimpse back into the fossil void because she very probably looks
much as her ancestors did from eight million years ago onwards.

CHAPTER 5

The first 'men'

As we have seen, the mysteries surrounding human prehistory are being steadily eroded by the torrent of fossil discoveries that are being made up and down Africa's Great Rift Valley. And the most recent finds tell us that the excitement of each discovery is by no means dimming: there are still many important things to learn. The recent discoveries span the whole length of the Rift Valley, and, appropriately enough, they span – and extend – the time period of human Pliocene and Pleistocene evolution for which we have real tangible evidence, not just hunches: from Tanzania and Ethiopia come remains or our primitive *Homo* ancestors who lived there three and a half million years ago; and in Kenya (at Koobi Fora) we have confirmation that *Homo erectus*, the evolutionary stage following *Homo habilis* and prior to *Homo sapiens*, really did arise in Africa.

About 20 miles west of Olduvai Gorge in northern Tanzania is an area known as Laetoli – it takes its name from the Masai word Laetoli, a red lily which grows in profusion there. The fossil-bearing deposits there, unlike those in the neighbouring Gorge and other notable sites, have been built up through the ages by wind-blown sand and volcanic ash: no lake or river sparkled its refreshing coolness here in Pliocene times. But water can't have been far away (there was the Olduvai lake, for instance, and probably streams too), because a profusion of animal and bird life left petrified remains that now are eroding out of the 500 feet of dry deposits: giraffes, rhinos, Deinotheria (elephant-like creatures with tusks drooping from their lower jaw rather than protruding from the sides of the trunk), monkeys, rodents, hyaenas – all thrived at Laetoli more than three million years ago. And so did *Homo*.

Mary Leakey and her colleagues have found lower jaws and some isolated teeth of about 13 individuals. They are definitely hominids, and they may well be primitive *Homo* – with such fragmentary offerings, it is difficult to be certain. These

63

creatures lived at Laetoli (or to be more precise, their bones finished up there) over three and a half million years ago.

And in one of the most tantalising of recent discoveries, it seems that these ancient inhabitants of Laetoli may – just may – have left their impressions for us *literally* in the sands of time. During a casual outing in one summer's expedition to the site, Andrew Hill, a British scientist working in Nairobi, happened to notice some strange fossilised shapes in what had been the edge of an ancient waterhole, a place where animals had come to slake their thirst more than three and a half million years ago. Jokingly, Hill called out to his companions, 'Look, I've found some fossilised footprints'. But it turned out to be no joke. They *were* fossilised footprints. Many of them, and of many different types of animals, most of which are now extinct.

It seems that ash from a nearby volcano settled in the shallow prints that had been impressed in firm but slightly damp mud close to the waterhole. The ash hardened into rock through the passing eons, preserving an extraordinary set of images from the past. When Mary Leakey was examining the prints she saw a line of six made by a single animal, and they were distinctly different from the rest. They were uncannily similar to modern human prints: they were rather broad for human footprints, it had to be admitted, but the toe pointed boldly forwards, quite unlike an ape's; and it certainly appeared as if the creature had strode upright as it passed the ancient pool. Could this be the first hominid prints of this great age? Could that line of prints have been made by an early form of *Homo*, a direct ancestor of modern man? It will be difficult ever to be certain, but scientists have already begun the frustrating task of analysing the biomechanics implied by the footprints in order to determine precisely how the creatures walked about, a study that will give us a fascinating glimpse into our history.

At about the same time that the treasures of Laetoli were being unearthed, Don Johanson and his colleagues almost a thousand miles further north at the Hadar sites in Ethiopia were proving just what a remarkable area it is: in one small locality they found, eroding out of a hillside, a whole series of bones which we believe are primitive *Homo*. Parts of skulls, lower jaws complete with teeth, arm bones, hand bones, leg bones, feet bones – all were there.

The fossil hominid bonanza began late one morning in

The sun sets over Lake Turkana.

The Koobi Fora spit where the fossil hunting team has camped ever since 1969. The spit provides a cool retreat in a sun-parched countryside.

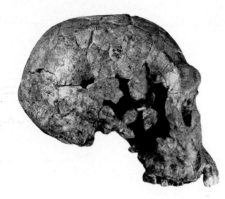

Australopithecus africanus (KNMER 1813)

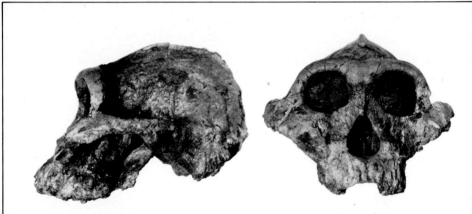

Australopithecus boisei (KNMER 406)

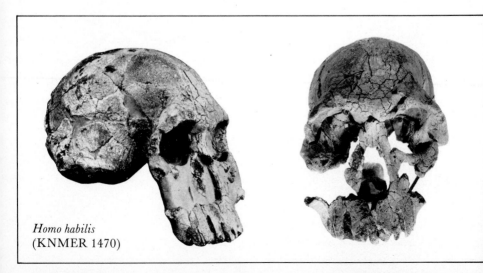

Homo habilis
(KNMER 1470)

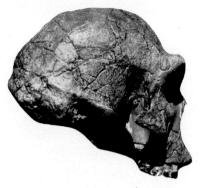

Homo erectus (KNMER 3733)

Left and above: crania of the four types of hominid found at East Turkana and in other parts of Africa. *A. africanus* lived at the same time as *A. boisei* but was smaller. *H. habilis* was ancestral to *H. erectus*.

Below: thigh bones of *(left to right) H. erectus,* modern *H. sapiens, A. boisei* and *A. africanus.* All the hominids who lived two million years ago stood and walked upright.

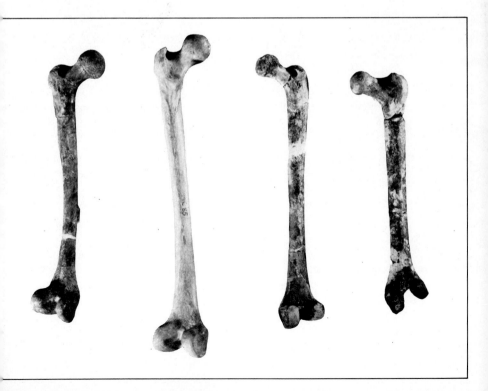

Richard Leakey examines a hominid jaw while seated in the dry bed of a seasonal stream. The occasional flash floods wash away the sandstone, revealing the fossils.

Richard Leakey excavates a *Homo erectus* skull at Koobi Fora. Kamoya Kimeu and Meave Leakey search for fragments.

Richard and Meave carefully remove sand from the fragile cranium. The cranium rests upside down in the deposits.

Meave clears a space around the cranium.

A tarpaulin gives shade as the excavation continues.

Richard applies a hardening liquid to prevent the fossil cranium flaking.

Richard uses a dental pick to remove particles of sand from the back and part of the top of the cranium.

The cranium is now top uppermost on a bed of sand in a container. Richard applies more hardening fluid.

Richard examines the sand-encrusted cranium.

Richard works at the cranium back at camp. The cranium, a fine specimen, is 1.3 million years old.

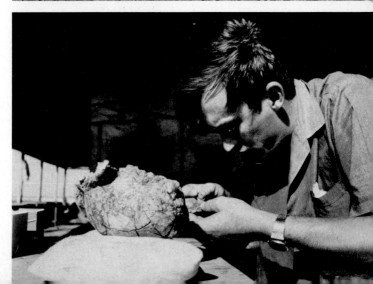

An aerial view of the East
Turkana fossil area looking
towards the lake. Note the
eroded sandstone terrain.

Richard Leakey walks
among the sandstone
sediments at Koobi Fora. A
desert rose and thorn
bushes grow from the
parched earth.

November 1975 when Mike Bush called to his colleague Tom Gray and said 'I think I've got something here'. It was Bush's very first day surveying for hominid fossils. Gray was hot, tired, and hungry, and he assumed that Bush had found a hippopotamus bone or something: 'that's what most people find on their first trip out'. With little enthusiasm Gray went over to where Bush was standing at the bottom of a steep gully: there, protruding from a block of sandstone were four hominid teeth. Bush *had* found a hominid. No hippo for him! Gray called Johanson to look at the teeth, and after some discussion they decided to excavate early next morning – it was too late in the day to start just then.

As it happened there was a French movie team and a National Geographic photographer at the site. They were delighted: the light would be better for photography in the morning. The excavation got under way the next morning with cameras whirring and shutters clicking, when Michele Cavillon, who was operating the field microphones for the movie crew, left the shade of a tree at the top of the excavation slope and went over to Tom Gray to show him some bones. She'd thought she'd found something. 'She was always bringing us bones,' Gray says, 'and usually they were nothing at all.' This time, however, was different. Together with some non-descript fossil and rock were two hominid ankle bones. Sound operator had turned fossil finder! A minute later someone wandered to the top of the slope and found a leg bone. Then Mike Bush found two more bones. 'Pretty soon everybody was rushing around finding them,' Gray recalls. Nothing like it had ever happened before in the history of palaeoanthropology.

Intensive surface collecting, plus a small excavation, eventually yielded several hundred bone and tooth specimens, all of which represents as many as 34 adults and 10 infants, one of whom was four years old when it died. Apart from the sheer spectacular nature of the discovery, scientifically the bone collection gives for the first time the opportunity of measuring the degree of variation of physical characteristics in a population of hominids. Previously we've had to compare individuals that died at different times and in different places. The Hadar group all died at the same time and in the same place.

In an adventurous excursion into skeletal reconstruction, Johanson managed to piece together a virtually complete hand from a collection of bones that undoubtedly once belonged to

65

several different individuals. To his amazement he discovered that, if his 'composite hand' didn't lie, these primitive members of *Homo* had hands very similar to ours. Whether they could move them as well as we can, whether they could achieve fine manipulative precision, we do not know. But we do now know that the structure of the hands was very advanced. And from their size, plus indications from other limb bones, it seems that, compared with the diminutive Lucy, these creatures stood really quite tall: probably close to five feet.

Apparently, close to three and a half million years ago, a group of these individuals perished in some sudden catastrophe. Very probably they were closely related to each other, forming a kind of proto-family group, a social pattern we might expect to see many times in human history. If so, what sudden catastrophe overtook them? Initially, it looked as if they may have been drowned by a flash flood in a stream bed, perhaps while they slept. But the geological nature of the deposits now rules this out. Were they victims of a particularly virulent disease? Had they chanced upon a luxurious abundance of poisonous berries or mushrooms and made the fatal error of eating them? Perhaps they were asphyxiated by noxious gases from the sudden blast of a volcano? We will almost certainly never know. Johanson says that 'I'm sure that as I get older and the years go by my interpretations will get more and more dramatic'.

Some people might suggest that Johanson's hominids are the victims of an ancient massacre inflicted by 'hostile neighbouring hordes' (to borrow a phrase from Lorenz). This is not impossible, of course, but the bones show no sign of unusual violence. But what motive makes one see a murder victim in every skeleton, and unbridled aggression in every prehistoric scenario? As a tortuous justification of today's bloody world perhaps? If real evidence of aggression exists in the prehistoric record, we will find it. But it is less than responsible to invoke ancient hostility where there is no *sound* reason to choose such an interpretation.

Meanwhile, as remains of our earliest ancestors were being uncovered in Tanzania and Ethiopia, fossil hunters at Koobi Fora were striking it rich too. There, during the past couple of years, remarkably fine specimens of so-called *Homo erectus* have turned up. This evolutionary progeny of *Homo habilis* lived in many parts of Europe and Asia until as late as half a million years ago. Then the basic *Homo sapiens* emerged, to be followed by truly modern humans perhaps 50 millennia ago (with a brief ill-fated excursion as Neanderthalers on the way). Until the *Homo erectus* skulls were discovered at Koobi Fora no one could be absolutely sure that this species of pre-human actually evolved in Africa. It always remained a possibility that *erectus* was born elsewhere, perhaps from a more primitive hominid that had migrated out of Africa, or (heretically!) from a hominid that originated outside the African continent.

One of the Koobi Fora *Homo erectus* skulls is around one and a half million years old (two others are a little younger), and interestingly it looks very much like *Homo erectus* that lived in China almost a million years later; this is Peking man.

Completing the jigsaw of human prehistory in East Africa is yet another very recently discovered skull from Koobi Fora, a skull that is roughly half a million years old. Although its appearance is not particularly dramatic, the cranium, which looks very much like *Homo erectus*, has distinct echoes of *Homo sapiens* about it, making it a very early member of our immediate ancestor, the basic grade of *Homo sapiens*. This skull, if it stands the tests of scientific scrutiny, would therefore be part of a unique sequence of pre-human fossil finds in a single area stretching over that crucial transition period of more than two million years ago to half a million years: *Homo habilis* to *Homo erectus* to *Homo sapiens*.

The name *Homo erectus* means, of course, erect man, a name that was coined many years ago. We now know that *Homo habilis* and almost certainly the primitive forms of *Homo* that lived at Laetoli and the Hadar (and presumably elsewhere too) also walked upright. Walking erect was therefore nothing special for *Homo erectus* to boast about – hominids had been doing it for eons. But what *was* special about him was that his kind had elaborated culture to such a degree that for the first time hominids could escape the climatic constraints of the tropics and live elsewhere as well: migration into cooler climes became possible. And, in the tradition of 'human' nature, anything that is possible is usually achieved: some time in excess of one million years ago bands of African-born *Homo erectus* made the journey into Asia and then on into Europe, carried, in all three continents, towards *Homo sapiens* by a powerful evolutionary momentum.

This special form of primate, who had been conceived in the

67

womb of opportunism, whose infancy was nurtured in a kaleidoscope of contrasting environments, and who, in rising maturity, invented a hunting and gathering economy, set out to hunt and gather elsewhere. Not to abandon its native Africa, but to colonise the rest of the world.

In spite of the relative sparsity of relics from human pre-history – the bones and stones – we can infer a good deal about how our forebears lived: how they walked; what they ate; how they developed their technology.

The same imaginative portraits that typically depict the australopithecines as being more hairy than their more human cousins also generally show our *Homo* ancestor striding out purposefully on two legs, with slouched lumbering australo-pithecines skulking in the background obviously incapable of a decent upright stance. We *do* have information on this, however, and it is sufficient to dispose of those *Homo* chauvinist notions of lumbering australopithecines. Indeed, our australopithecine cousins may possibly have been better walkers than *Homo habilis* two million years ago!

Instead of relying on guesses garnished with a good dollop of prejudice as a way of showing how the early hominids got themselves about, Owen Lovejoy, an anthropologist from Kent State University, has been indulging in biomechanics. He has some surprising and important things to say about hominid locomotion.

For reasons that still leave us groping in the dark, the pressures of natural selection invented upright walking in our hominid ancestors some time between 15 million and about three million years ago. This style of locomotion is therefore a very old invention. There are many good reasons for getting around on four legs rather than two: it is cheaper energetically for a start; and the stresses and strains imposed by the body and its various antics are distributed among four joints rather than two. But once you stand on two legs, not only have two joints to take the body weight instead of four, but the mechanics change dramatically too.

Two major muscle blocks control our legs: the quadriceps at the front of the thigh which straightens the leg; and the hamstrings at the back, which bend it. In four-legged creatures the hamstring muscles do the important work during the so-called power stroke of the leg, the one that moves the body along. In us it is the quadriceps. That is one change.

The others allow us to walk along evenly rather than waddle like a duck – or a chimpanzee, for that matter, during its brief excursions into bipedalism. First, the angle between the knee and the thigh bone (the femur) has to be big enough to make sure that we place our feet more or less under the centre of our body when we walk. You can check this by looking at your tracks in the snow or sand: they follow you in a straight line. If, however, you were a chimpanzee then you'd see two lines of footprints: the chimp gets the weight of his body over, say, the right foot by moving his body in that direction, and then over the left by shifting it back again – hence the waddle. Humans make sure that the feet are in the middle to start with.

Because of the significant angle between the knee joint and the thigh bone there is a very real danger that each time the quadriceps contract the knee cap (to which the muscles are attached) will be ripped out of the side of the leg. Fortunately, nature intervened to prevent this catastrophe, and there is a ridge of bone (called the lateral condyle) on the lower part of the thigh bone which stops the knee cap slipping.

Not only that, but each time we take a stride there is imminent danger of the body collapsing, or at least seriously sagging, on the side with the leg off the ground. Again nature intervenes, and the sag is averted by a couple of muscles attached to the upper part of the thigh bone and the pelvis which hoist the body up as we walk.

Lovejoy has been analysing these structures and their detailed movements, both in modern bones and in a three million, three hundred thousand-year-old fossil that has come to be known as Johanson's knee. Don Johanson, the anthropologist from the Cleveland Museum, Ohio, has two knees of his own, of course, but the one Lovejoy has been working with was found at the extraordinarily fertile fossil site near the Awash River in Ethiopia. Late in the afternoon of the last day of October 1973, the second year of the joint American/French expedition to the site, Johanson was searching in the sun-baked sediments of an ancient lake when he spotted a small shin bone eroding from the ground. This was the first hominid fossil the team had come across at the site, and Johanson's inevitable excitement intensified when, just a few feet away in a small gully he saw the lower end of a thigh bone: the two bones fitted to make a perfect knee, without doubt from the same long-dead individual. Johanson can't be certain what type of hominid it belonged to, but with an age of almost three and a half million years, it is safe to say that it is 'primitive'.

One of the most important qualities of Johanson's knee is its remarkable state of preservation: the 'bone' is in such good condition that Lovejoy was able to put it through its analytical paces, much as if it were a modern bone. The first thing that Lovejoy noticed was that the angle between the knee and the shaft of the thigh bone was significant, certainly within the modern human range. This firm indication of well-developed upright walking was then confirmed by the prominent lateral condyle, the lip of bone that prevents repeated dislocation of the knee cap. The structure was certainly 'modern', and so, as it turned out, were the biomechanics. In every test that Lovejoy invented to determine whether the ancient knee had belonged to an upright individual, the answer was positive, even for the most sophisticated and exacting analysis.

The implication of Johanson's knee, then, is that at least one type of hominid was walking around the Pliocene landscape with a perfectly modern upright posture. Unfortunately, there are not enough superbly preserved knees in the fossil record to allow each type of hominid to be tested separately. Instead, we can look at the upper part of the thigh bone and the pelvis. Now, there is no doubt that these bones of the australopithecines are different from those of *Homo*, and certainly from modern humans: in the australopithecines the pelvis is narrower and longer; the ball of the femur is smaller; and the neck of the femur (which attaches the ball to the shaft of the thigh bone) is strangely flattened. In the past many people have interpreted these differences as indicative of a slouching gait in the australopithecines, with no better reason than that the bones are different from *Homo*'s – more *Homo* chauvinism here.

Lovejoy has been looking into these problems too, and he enjoys being heretical by suggesting that the differences in the pelvic and leg bones could actually mean that it was *Homo* who had a hard time walking upright two or three million years ago, and even that, if australopithecines were with us today, then, biomechanically, they would be more efficient walkers.

Homo's pelvis and upper thigh bones are different from the australopithecines', Lovejoy suggests, because our ancestors had bigger heads. That may sound odd, but what he means is that the birth canal in the pelvis had to be broadened so that the large-brained babies could emerge into the world relatively unscathed: even so, during birth the ligaments on the pelvic bones have to loosen so as to widen the pelvis even more, and the baby's head is usually squeezed into an almost pointed shape as it makes its hazardous journey from the womb to the outside world.

Although men don't give birth to babies, they too have relatively wide hips, though generally not as wide as women's. Lovejoy calculates that this difference means that the hip joints in men suffer between 20 and 30 per cent less stress than they do in women. It is conceivable that with even narrower hips the australopithecine joints were less stressed still. The long neck of the femur and the smaller femoral head may simply be adaptations to reduced biomechanical demands in australopithecines' hip joints.

Differences in *anatomy* do not necessarily imply important differences in *function*, however: the australopithecine hip joint may be more *efficient* than *Homo*'s in terms of the stresses and strains and energy involved, but they may not have been more *effective* in propelling their owners in an upright posture. The most charitable view, until we know for certain one way or the other, is that all the hominids stood upright and walked with a respectable striding gait.

These discoveries imply that getting around on two legs is an extremely ancient hominid practice. Probably it stretches back out of the Pliocene and into the fossil void of the Miocene. It may be as old as eight million years. This is remarkable, and it certainly takes our view of prehuman adaptations much further back into our history than anyone would have guessed even a couple of years ago. When we view human prehistory we usually think of proto-human activities such as creating stone tool industries and perhaps occasional organised hunting as part of a food-sharing economy. These types of behaviour were undoubtedly influential in the emergence of mankind, but they almost certainly played little part in encouraging the earliest hominids to stand on two legs instead of four.

Food is a great preoccupation of most primates, and the hominids two to three million years ago were unlikely to be exceptions. What did they eat? The 'traditional' view is that our *Homo* ancestors ate a mixed diet of plant material and meat (both hunted and scavenged); *Australopithecus boisei* is slotted into a strictly vegetarian role; and *Australopithecus africanus* comes somewhere in between, somewhat like modern olive baboons which forage opportunistically for roots, shoots, insects, and occasionally for small animals. This view is vague enough so that it almost cannot be false; but it is not very helpful.

What do we have to go on? Teeth and stone tools. The occasional occurrence of deliberately shaped stone tools with the bones of animals tells us that at least two million years ago, possibly three, one of the hominids was going in for meat-eating in a way that no other primate does. We assume that the tool maker was *Homo*, and we postulate the emergence of a mixed economy in which plant and animal foods are shared between males and females. This is as much a social revolution as a dietary one, and it was certainly crucial in the emergence of mankind. For the moment, however, we must put that aside and come back to the teeth.

First of all, the molar teeth of all hominids are large relative to the size of the body. And the layer of hard enamel on the outside is unusually thick too, certainly thicker than modern apes'. We can stick our neck out and say that during the course of the long human career, plant foods have been a tremendously important item on the prehistoric menu, for *Homo* as well as for the australopithecines. The question is, what sort of plants? There is a wealth to choose from: nuts, fruit, leaves, seeds, stalks, roots, buds, and grass.

Most of the large primates, while specialising on one general type of plant food (fruits, or leaves, or hard roots and seeds), are opportunists: birds' eggs, lizards, grubs, insects, birds, small mammals, all are added to the diet when the opportunity arises. And we can expect that the early hominids were equally opportunistic. But what were their specialities?

A glance into the jaw of *Australopithecus boisei* is enough to convince you that those enormous teeth must have imparted gargantuan pressures for crushing even the toughest of food morsels. And when you learn that the jaw was driven in rhythmic grinding motions by more than three pounds' weight of muscle, your conviction would be strengthened. But you would almost certainly be wrong. In some brilliant calculations Alan Walker has come up with the surprising answer that in spite of the size of the teeth, in spite of the dimensions of the jaw, and in spite of the massive amount of muscle that drives it all, *Australopithecus boisei* generated only the same amount of pressure between its grinding teeth as can modern humans. The size is impressive, but the performance is not.

The reason *Australopithecus boisei* has such massive jaws is that it is a big animal: he needed to process a great deal of fodder to fuel his large bulk. The cavernous mouth lined with millstone molars surely aided this great fellow in consuming an enormous bulk of food throughout each day. (It would be amusing if 'Nutcracker man' turned out instead to be 'Grass-eating man'!)

The big australopithecine may have preferred to feed in relatively open grassland, eating not just the grass, but seeds and roots, and insects too. And it may be that his smaller cousin stuck closer to the cover of bushes and trees (*africanus* would certainly have been more vulnerable to predators than *boisei*). More research on fine details of the teeth should give the answer.

Both species almost certainly went about their daily business in social groups, but within the groups each individual would have fed itself, just as chimps, gorillas, baboons, and all other primates do today: socialising, but not social eating. At night the foraging groups of australopithecines would make their way back to the safety of their sleeping trees or cliff ledges; again, just as modern chimps and baboons do.

Meanwhile, life among our emerging *Homo* ancestors was probably very different.

The most tangible evidence of this, and indeed the most tangible items of all in the prehistoric record, are stone tools. Hominids started making stone tools to a purposeful and organised pattern at least two and a half million years ago, and probably earlier. What's more, occasionally at least, they made and used the tools in specific places, rather than habitually wandering around the countryside stone-knapping with gay abandon. In other words, we have evidence for ancient camp sites. And as we scan the period from three million years down to one million (by which time some of our ancestors had begun to colonise Europe and Asia), we see a steady increase in the number of camps and in the complexity of the tool technology discarded in them. These mute remains are the products of tremendous social and intellectual revolution in the lives of the early hominids as they travelled the road to modern humanity.

If we cast our minds back to the two-million-year-old KBS camp site by Lake Turkana (the one featured in our imaginative reconstruction in chapter 1) it soon becomes apparent to us that stone tools were more than just the casual products of idle hands: the hominids on the KBS camp site had to travel several miles to obtain the lava cobbles from which to make the tools, and this must have demanded considerable planning and foresight as well as physical effort. The technology, however, was crude: tennis-ball-sized pebbles with four or five flakes knocked off to make a chopper; an angular cobble broken in half to make a so-called discoid; blade-like flakes; small angular flakes, and so-called polyhedrans. We can only guess how our ancestors used these tools in their daily lives: they would have to slice meat occasionally, perhaps using the small or blade-like flakes; digging sticks must be sharpened, a job easily achieved with a freshly struck flake; discoids may have been used to scrape animals' skins or to soften bark; nuts must be pounded and bones broken to release the succulent marrow – any piece of heavy cobble might serve here.

It is probably more useful to try to think of as many jobs that are possible with any particular type of tool, than to attempt to pin specific shapes to specific functions: this is a pastime anyone can indulge in. Certainly, one important group of items that represents a frustrating void in all the fossil record is plant materials: plant foods and wooden implements usually disappear without trace. Many of the stone tools might have been used for making more elaborate implements from wood, or perhaps for preparing vegetable foods. All too often this is ignored in scenarios of pre-human life that drip with slaughtered animal corpses (and other hominids too) and jangle with ugly-looking stone weapons.

If it were hefty stone weapons that the hominids of the lower Omo valley lusted after around two and a half million years ago, then they seem to have been disappointed. All the early stone artefacts that so far have been uncovered in this area are tiny fragments of milky-veined quartz, usually about an inch across. Our ancestors in this area appear to have had their stone tool technology severely limited by the absence of basic materials in the riverside areas in which they chose to live. And the rocks they were able to find, which are probably tumbled along in streams from some mountains 15 miles distant, were of quartz, which is simply not suitable for manufacturing large, sharp implements.

At first sight the quartz flakes look so unprepossessing that they were dismissed as implements: they look more like debris. But a close examination of the pieces shows that the

74

hominids retouched them to produce sharp edges. And there are signs of use on them too, showing that they were employed to cut some hard materials. The tools at the lower Omo tell us that when we see a stone culture, at least part of the 'design' might be a consequence of the physics of the raw materials employed.

So far the oldest man-made stone artefacts have been found at the Hadar by the American/French joint project. They were made by hominids living somewhat more than two and a half million years ago. The tool-kit is simple, but its discovery was dramatic in the context of the search for human origins. A young French archaeologist, Helene Roche, felt sure that one day she'd find the tools that she believed the ancient Hadar hominids (probably *Homo*) must have been making. For several years she searched and searched, but although she found some fairly recent stone implements, the really old tool-kits eluded her. Then one morning, in the fall of 1976, while she was out on a routine survey, she happened to glance at the side of a small gully, and there, eroding from the sand were tools. They were crude scrapers and choppers made from lava cobbles – but they were unmistakably the products of ancient 'human' hands. To discover just how long ago those hands had struck the tools from stream-smoothed cobbles, Roche had to consult the geologists. She was ecstatic when she was told that they were at least two and a half million years old, making them the oldest artefacts so far discovered. Not content with that, she believes that one day she'll find tools that will stretch the archaeological record still further!

Out of all the archaeological sites in East Africa, however, the one with the most splendid collection is undoubtedly Olduvai Gorge in Tanzania. There during much of the past two million years a fluctuating closed lake, sometimes measuring 20 miles north to south and just less than 15 miles east to west, slowly evaporated in the heat of the tropical sun. The skyline to the southeast was dominated by the volcanic highlands of Ngorongoro, Lemagrut, and Olmoti. Grassland savanna stretched from the lake's mudflats to the forested hills in the north, east, and south. Streams feeding the lake ran down from the hills, mainly from the east. The lakeside was rich in plant and animal life. Now the waters have gone, and for the most part it is a dry corner of the Serengeti plain. A recent seasonal river has cut through the ancient sediments, excavating a 300-feet deep gorge that exposes a remarkable slice of human prehistory.

In more than 30 years of patient and detailed work at the Gorge, Mary Leakey has documented the development of stone tool technology over much of the past two million years. The sequence is remarkable. The hominids living at the lakeside almost two million years ago made and used a crude stone tool technology very similar to the one with which the KBS site hominids made their living. Composed roughly of primitive choppers, scrapers, and flakes, this technology has been called 'Oldowan'. The Oldowan design continued to be the basic toolkit of the Olduvai hominids for a million years, a staggering continuity! The stone tool culture didn't remain static for the whole of that time, however. It gradually became more complex, so that by one and a half million years ago there were more than 10 identifiable implements in the tool-kit compared with the six in the basic Oldowan. Mary Leakey named the 'new' technology 'Developed Oldowan'.

Interestingly, by the time Developed Oldowan tool-kits were being knapped out at Olduvai, a similar though not identical technology was emerging at Lake Turkana. This technology, known as the Karari industry, may have been elaborated by the descendants of the hominids who lived a million years earlier at the KBS camp site. Technological progress during this era was not of the pace to which modern business executives have grown used, but this does not imply that social and behavioural progress was equally tardy: technological progress *was* unhurried, but perhaps in those days prizes were given for inventing new ways of using old tools rather than for developing new kinds of tools!

Around one and a half million years ago an interesting phenomenon occurred at Olduvai: a second stone tool culture arrived and began a long coexistence with the indigenous Developed Oldowan tool-kit. The new technology, known as Acheulian, is best known for its so-called hand axes: these are carefully fashioned teardrop-shaped implements for which, embarrassingly, no one can think of a good use. Some of them are so heavy as almost to defy any possible practical use for them at all; while others you can hold comfortably in the palm of your hand and they may well have been put to the use for which their name implies. Perhaps they were simply the way a stone-tool knapper demonstrated his skill: a kind of prehistoric trade mark!

The Acheulian industry continued, with steady refinement,

right through to the end of the sequence at Olduvai. And, clearly, when some African hominids migrated north, they took their technology with them: Acheulian tools were used in Europe as late as 200,000 years ago. Meanwhile our ancestors in Asia manufactured stone tools very much like the Developed Oldowan, but the industry did not survive as long as the Acheulian. It was the Acheulian that finally predominated, until it too was overtaken by more advanced stone technologies.

What was going on when the Olduvai hominids were manufacturing the two distinct technologies at the same time, and over a period of half a million years? Were they made by two different hominids, the developed Oldowan by a remnant population of *Homo habilis*, and the Acheulian by *Homo erectus* people who migrated into the area, finally to oust the indigents? Perhaps two different tribes of the same type of hominid used different design in stone tool industry in order to express their identity, just as technologically primitive people do today; one culture eventually dominating the other. Maybe the Oldowan tool-kit was best for one set of jobs and the Acheulian for another, both being used by the lakeside hominids for the different purposes for which they were designed. We simply don't know. But it is certainly another situation in which one should play the game of inventing as many possibilities as one can, rather than be dogmatic about any particular fancy.

Humans, as we all know from experience, are inventive creatures: we improvise when the situation demands it. Caught with a screw to tighten, but having no screwdriver to hand, we look around for a substitute: a knife, a coin, a key, a scrap of metal, even a thumb nail. Although our hominid ancestors would not have had this particular problem to cope with, their daily lives must have been full of practical tasks for which they had to improvise. For instance, the man who, in our hypothetical scenario, dug a water hole with a stick might well have used the shoulder blade of a small antelope, had one been lying close by. And it would be truly remarkable if, in the prehistoric record, no animal canine found itself serving as a knife after it had left its owner's jaw. Indeed, all we have to do is look to the Yanomamo indians of South America to see a present-day example: these people use the lower jaw of a wild pig as a plane for shaping wooden implements; the projecting canine tooth is extremely sharp and cuts off slivers of wood very easily.

77

Unlike Raymond Dart, however, we do not suggest the past existence of specific bone tool cultures, merely that bones, teeth, and horns were probably put to whatever opportunistic use they were most suited.

Dart invented the notion of an 'osteodontokeratic' culture (meaning bone, teeth, and horn) after he had analysed the animal bones that had been hacked out of the Makapansgat limestone caves. He was very impressed by the fact that bones from some parts of the skeleton were more common than others in the cave deposits. In 1957 he wrote 'the disappearance of tails was probably due to their use as signals and whips in hunting outside the cavern. Caudal and other vertebrae may also have disappeared because of the potential value of their bodies as projectiles and of their processes (when present) as levers and points. The femora and tibiae [leg bones] would be the heaviest clubs to use outside the cavern; that is probably why these bones are the least common. Humeri [upper arm bones] are the commonest of the long bones; probably because they would be most convenient clubs for the women-folk and children to use at home.'

Dart envisaged the hominids living in the caves and basing their technology on nature's bony offerings. Yes, nature's bony offerings were undoubtedly put to good use from time to time, but not, as it turns out, as the basis for an 'osteodontokeratic' culture at Makapansgat. In a typically thorough piece of research Bob Brain, working at the Transvaal Museum in Pretoria, discovered that, when exposed to the elements and to the gnawing attentions of scavengers, some bones survive while others disintegrate. It all depends on the density and the thickness of the bone. Not surprisingly, then, the survivability of different types of bone (arm bones, leg bones, jaws, vertebrae, and so on) matches very closely the accumulation pattern of fossilised bones in the Makapansgat caves: the tough ones finished up in the caves, while the weaker ones disintegrated or were eaten before they had a chance of becoming part of the bone and rock conglomeration. The caves were probably rarely, if ever, used as a hominid home.

The osteodontokeratic 'culture' therefore turns out to be no such thing: it was simply a consequence of the fact that scavengers can crunch up some bones whereas they find others impossible to eat, with the natural elements playing their part too. This is a cautionary tale, and one that should be remembered by anyone in search of meanings in 'significant patterns' buried among the prehistoric record.

Incidentally, as part of his long and careful study into the fate of corpses, Brain discovered that, compared with many of the bones of, say, antelopes, primate skeletons present very little problem for carnivores: they can chew their way through most of a primate skeleton, apart from the cranium and the lower jaw. Brain's ingenious experiments were on modern primates and modern carnivores, but the same rules undoubtedly applied to the ancient hominids, thus explaining why in the bony litter of the past, skulls and jaws are most common. And most common of all are lower jaws of robust australopithecines: these must be the toughest pieces of hominid architecture in all of prehistory.

With the evidence of the bones and stones now proving solid support behind us, we can focus more specifically – and with more than a mere thread of confidence – on our ancestors' way of life: on their economy, their intellectual and social expansion, and, last of all, the gradual embracing of warfare as an extreme implement of interaction.

CHAPTER 6

An ancient way of life

When the concept of agriculture first arose about 10,000 years ago it precipitated the decline – slow at first, then more rapid as time passed – of a hunting and gathering existence that had dominated human history for at least two million years, and possibly much longer. The forces of evolution that, through the late Pliocene and the Pleistocene epochs, moulded the human mind and shaped our psychology and our social responsiveness are those embedded in the hunting and gathering way of life. So much so that today we look out on a technologically sophisticated and socially divided world with the brains of hunter-gatherers in our heads.

If we wish to understand the nature of human nature then we would be foolish to ignore the lessons that contemporary technologically simple societies can teach us. Not that present-day hunters and gatherers – of which there are now just a few hundred thousand in a world population of more than four thousand million – are 'fossilised' societies frozen in an age-old and primitive way of life. They are modern human beings pursuing a lifestyle that has its roots deep in the origins of humanity.

What, then, are the questions we can ask of these people? What threads of humanity can we hope to unpick from their complex way of life?

We want to know what factors influence the social organisation of hunting and gathering communities; what values these people espouse; what skills, intellectual and social, do they need in order to survive in the hunting and gathering economy; and what relationships do they have with the world around them?

When American anthropologist Napoleon Chagnon crouched low and crawled through the small entrance to a Yanomamo village more than 10 years ago, these were the kinds of questions that were in his mind. The Yanomamo Indians, whose homes straddle the densely forested border between southern Venezuela and northern Brazil, make their living by a mixture of hunting and gathering, and cultivation: they hunt a wide range of animals, from lizards to monkeys (usually with only modest success), and they collect wild mushrooms, roots, nuts, and fruits of the forest; but for the most part they depend on plaintains and other vegetables that they grow in the village 'garden'. Chagnon knew little of the Yanomamo, the people with whom he was eventually to live for more than 18 months, and his first encounter with them was a great shock. Filled with the thrill of meeting for the first time the people he had chosen to study as an anthropologist, Chagnon crawled through the low entrance to the Yanomamo village and was confronted by 'a dozen burly, naked, filthy, hideous men staring at us down the shafts of their drawn arrows! Immense wads of green tobacco were stuck between their lower teeth and lips making them look even more hideous, and strands of dark green slime dripped or hung from their noses. . . . I was horrified. What sort of welcome was this . . .?' Later he dubbed them 'the fierce people', because of their devotion to aggression as a major element of their culture.

Meanwhile, as Chagnon was attempting to understand the cultural imperatives of the Yanomamo, a group of researchers was studying the social life and subsistence economy of a true hunter-gatherer people, the !Kung San,* in northern Botswana. Here, in the sun-parched hills of an area known as the Dobe, which is on the northern fringes of the Kalahari Desert, the !Kung make a living on what nature has to offer: they hunt animals (again, with less than spectacular success); they collect nuts and roots and other vegetables; but they cultivate nothing, excepting, perhaps, leisure. Because of the careful studies organised by anthropologists Richard Lee and Irven DeVore, the !Kung have become something of a model for what true hunting and gathering is all about. And for reasons that are obvious when you look at them, the !Kung have been called 'the gentle people'.

Because 'the gentle people' are true hunter-gatherers, while

* The !Kung San, and their neighbours the G/wi, speak a so-called click language. The ! is an alveolar-palatal click produced by firstly pressing the tongue firmly against the roof of the mouth and then sharply snapping it down. A loud pop results. The / is a dental click produced by first placing the tongue against the back of the upper front teeth; it is then sharply released, giving a reproving 'tsk' sound.

the 'fierce people' supplement their hunting and gathering with limited agriculture, some of the comparisons one can make between the two societies are pertinent to the economic and cultural transition our ancestors initiated 10,000 years ago when they turned away gradually from the hunter-gatherer way of life and took up an agricultural existence. We want in the final chapter to point up some of the contrasts in these two different lifestyles and in the psychological attitudes that flow from them. But here we're more interested in the commonalities of technologically simple subsistence.

The traditional view of hunter-gatherer societies is that they are organised around what anthropologists have been pleased to call 'patrilocal bands'. There are three basic ingredients in such bands: first, everyone must marry someone from outside the group; second, when women marry they move to their husband's band; and third, the band maintains a strictly marked-out, assiduously defended territory. The overall impression is of a rigid honeycomb of male-dominated, socially stable bands, or hordes as they are sometimes called, each band surviving on the resources available in its allotted territory. It sounds very elegant, very neat.

But, as with most 'traditional' views, this idea pays greater service to elegance and to theoretical convenience than it does to the truth. Although some elements of the patrilocal band do apply to hunter-gatherer communities, the notion as stated is too rigid, too inflexible. It is now clear that the vast majority of hunter-gatherers live in relatively open social groups, and that neighbouring bands share overlapping, changeable territories. This pattern holds for hunter-gatherers in every part of the globe, from the scorching deserts of southern Africa to the frozen wastes of the Arctic. And for good reason. The flexibility of the bands themselves and of the territory they exploit has important social and economic advantages: social tensions can readily be defused; and localised food shortages and surpluses can be accommodated with benefit to many bands sharing overlapping regions. In the jargon phrase of biologists, flexibility is socially and ecologically adaptive.

The network that holds the people together, both within bands and between them, is kinship. And the predominant behaviours that make the whole system run smoothly are sharing and cooperation. These factors apply as much to the Yanomamo as they do to the !Kung, in spite of their very different cultural values – aggression on the one hand, and gentleness on the other.

The !Kung San are part of a people who have lived in southern Africa for at least 10,000 years, possibly as long as 50,000. And as far as one can tell the !Kung way of life today is very similar to their ancestors' many millennia ago. The !Kung and their neighbours make a living in the demanding lands of the Kalahari Desert, not because they have been forced into marginal territory by European colonisation, but because they always have. However, before the 1650s when the Dutch settlers arrived, the San occupied the whole of southern Africa, from the Zambezi Valley to the Cape of Good Hope. The Dutch invaders eventually wiped out most of the San, apart from those living in the Kalahari Desert where it straddles parts of Botswana, Namibia (illegally occupied by South Africa), and Angola. It was the Dutch who gave the San the derogatory name Bushmen, a term that should no longer be used.

Inevitably, water – or rather the scarcity of it – dominates much of !Kung life. (Availability of water must have been something of a preoccupation for the early hominids too, not necessarily because they sweated it out in parched terrain like the Kalahari Desert, but because they probably had only limited means of transporting liquids.) Rain never falls in the northern Kalahari between June and September, and during this time the !Kung congregate in relatively large camps around the eight permanent water holes in the Dobe area. In the hot summer months of October to May the sparse rains fill up temporary water holes, such as shallow depressions between the dune-hills, and hollows in mongongo nut trees. It is when the otherwise dry landscape sparkles here and there with temporary water holes, that the !Kung disperse into smaller bands of about 25 individuals who camp for a few weeks at a time near one of the ephemeral oases. Contrary to popular mythology, the bands are not driven from place to place by the unrelenting spectre of starvation and deprivation: hunter-gatherers move their camps regularly to exploit fresh food resources elsewhere (usually very close by) – and it is their way of life.

As an interesting contrast, the G/wi San, who live almost 300 miles southeast of the Dobe area in an even more arid part of the Kalahari, congregate and disperse in precisely the opposite pattern from the northern cousins: they come together during the wet season (while the !Kung are dispersing) and separate again when the rains finish (just as the !Kung are congregating). The *ecological* reason for the G/wi's behaviour is that, unlike the !Kung, they have no *permanent* water holes in their area and they spend much of the year in small groups, their only source of water being several different types of bitter melons and other succulent fruit. (The G/wi are probably the only people on the earth who survive for so long without free-standing water.) In the brief rainy season temporary water holes fill up for about six to eight weeks, and it is on these that the G/wi converge and form larger bands.

The *social* reasons for these fluxes in group structure are several: first, the congregation allows much socialisation, which everyone clearly enjoys greatly, but which also re-establishes contact between people who might otherwise be separated; and second, during the flux in the band structure, its composition can alter too – bands may split, small ones coalesce, families move from one band to another. It is this dynamic state of the G/wi groups that appears to be a crucial mechanism for defusing social tensions by allowing the dissidents to separate from each other. And if one family would rather be with another one, simply because they enjoy the other's company, they can do that too.

In any case the need for band flexibility is obviously deeply ingrained in the minds of hunter-gatherers, to such a degree that they seem capable of deluding themselves in order to achieve it. For instance, Colin Turnbull tells of hunting groups among the Mbuti Pygmies in the Ituri forest of the Congo. One type of group hunts with nets, an activity that demands the co-operation of perhaps six or more families, so the bands are large. When the short honey season arrives – providing a delicacy that virtually all hunting and gathering people relish, consuming, as they do, not only the golden yellow liquid, but the waxy honeycomb and the squirming larvae as well – the bands split up into small units in search of honey: they say that at this time the hunting is so good that animals can easily be caught by hand. When the honey feasts are finished they reform into net-hunting bands, being very careful to ensure that any antagonisms of the previous hunting season are avoided by the simple expedient of the antagonists joining different bands: there is no formal restructuring; the mechanism operates tacitly.

Meanwhile, the bow and arrow hunting Mbuti do precisely the reverse in precisely the same situation and give precisely the opposite reason. Hunting with bows and arrows demands small numbers: the Mbuti men like to hunt in groups of three because, they say, they get the best results this way. But when the honey season arrives they claim that the game is so scarce that maximum cooperation is needed to ensure any success at all: so they form large bands. The same possibilities arise for a period of socialising and for soothing social tensions by reorganising themselves. And, as far as Turnbull could determine, the hunting opportunities really did not seem to alter much, if at all, during the honey season!

It was the eighteenth-century philosopher Thomas Hobbes who described the lives of primitive people as 'Nasty, brutish, and short'. Hobbes' phrase also aptly sums up many people's vision of our ancestors' physical appearance. We now know through the fossil discoveries of recent years that the latter impression is wrong. And so, it appears, was Hobbes.

As with the vast majority of hunter-gatherers, the !Kung divide their economic activity so that the men hunt while the women gather nuts, roots, and other vegetables, depending on what are the tastiest foods in season. On average adults work for between 12 and 19 hours a week, a devotion to the food quest that can hardly be termed excessive! Although girls may begin an adult life at around 15 years of age, boys commonly do not step into the adult world until they are at least 20. And by the time people reach 60 they generally 'retire' and are then cared for, respected, and fed for the rest of their days: the old are greatly valued for their experience and wisdom. Childhood and old age are therefore free of stress and obligation in the !Kung society.

What kind of society is it, then, where working life begins at 15 years at the earliest, and finishes at 60, with an average of about two and a half hours labour each day in between? American anthropologist Marshall Sahlins describes it as the original affluent society, where finite needs are satisfied with a minimum of effort. Certainly it doesn't appear to be a recipe for an existence that is nasty, brutish, and short.

Admittedly, the !Kung are particularly lucky in having a virtually unending supply of mongongo nuts which grow in groves on the tops of the hill-dunes. Mongogo nuts are staple food for the !Kung: on average they eat 300 nuts each day, which contain calories equivalent to two and a half pounds of cooked rice, and the same amount of protein as in 14 ounces of beef! No wonder one of the !Kung said to Richard Lee 'Why should we plant when there are so many mongongo nuts in the world?': each year they collect thousands of pounds of the nuts, but many thousands more lie decaying on the ground.

Are the !Kung a misleading example of life as hunter-gatherers, as there are so many mongongo nuts in their world? Apparently not. For two reasons. First, when Lee was making his measurements about the amount of work the !Kung did it was the dry season of the third year of the most severe drought ever recorded in that part of the Kalahari. Nearby in Botswana more than a third of the population of 500,000 Bantu pastoralists and agriculturalists were in such a poor state that the United Nations World Food Programme had to launch a famine relief exercise to save them from starvation. Meanwhile, the natural products of the land were maintaining the !Kung in excellent health and with very little effort on their part. If Lee had been with the !Kung in a normal rainfall year he may well have found them spending even less time gathering food, leaving even more free time for leisure, socialising, and dancing.

A second reason for believing that, although the !Kung may be unusual in having such a stable supply of a particularly nutritious single item of food they are not unusual in having so much leisure, is the example of the G/wi San in the cruelly marginal lands of the Central Kalahari. Surely these people must be clinging grimly to the very edge of a pitiful existence, having to dig not only for food, but for water as well? As it turns out the G/wi do have to devote more time and energy to the food quest than do their !Kung cousins: on average adult G/wi work for just less than 33 hours a week – but this is still less than the average working week of people turning the wheels of technologically sophisticated economies of most affluent western countries.

Truly objective scientific assessments of the labour of hunter-gatherers in many different environments most often come up with figures somewhere in between those for the !Kung and the G/wi. And although the majority of the pioneering Victorian explorers confirmed for themselves the prejudices embodied in Hobbes' thoughts, there were one or two who were more perceptive. For instance, in 1841 Sir George Gray wrote of the Australian aborigines: 'In ordinary seasons they can obtain in two or three hours a sufficient supply of food for the day'. In a tone espousing the Victorian work ethic, he went on to say, however, that 'their usual custom is to roam indolently from spot to spot, lazily collecting it as they wander along.' Speaking about the same people in 1845 John Edward Eyre said 'I have found that the natives could usually, in three or four hours, procure as much food as would last for the day, and that without fatigue or labour.'

When, in 1952, Melville Herskovits wrote in his 'Economic Anthropology' that the San of the Kalahari and the native Australians were 'a classic illustration of a people whose economic resources are of the scantiest . . . so that . . . only the most intense application makes survival possible', he himself was a classic illustration of a people who applied myopic and biased economic analysis to a way of life they did not understand. And it was this kind of analysis that nurtured the conclusion that the apparent absence of elaborate culture among hunter-gatherers was a consequence of their weary devotion to an unrelenting food quest. Elaborately carved and beautifully decorated accoutrements of culture do not play an important part in the lives of hunter-gatherers, not because they don't have culture, but because their whole lives are geared to possessions that can readily be carried from one camp to the next (or rather it is more true to say that a hunter limits his possessions to things that his *wife* (or wives) can carry to the next camp). Hunter-gatherers prefer to carry their culture in their heads rather than transport the objects of culture on their backs: myths, songs, story-telling, and dances, all are part of their rich cultural fabric.

One salutary example of how culture may be moulded around ecological necessity comes from the G/wi San and their complex series of taboos on eating certain kinds of meat. G/wi culture prohibits hyaenas, vultures, hunting dogs, and lions from their menu, as do many hunting and gathering people. But the most interesting taboos concern the steenbok, spring-hare, kori bustard, and tortoise. Only infants and people over 40 years of age may eat kori bustard and tortoise. And steenbok and spring-hare may not be eaten by married couples with infants under six months old, nor by people who are able to fall into a trance in the gemsbok dance. By contrast, everyone joins in the feast when, on the infrequent occasions someone kills eland, kudu, gemsbok, hartebeest, wildebeest, giraffe,

87

springbok or ostrich. The difference between this group of animals and the tabooed ones is size: the tabooed meat comes from small animals that simply cannot feed a whole camp, whereas the others are large and can provide a meal for all.

The result of the taboos is that the meat from these small animals (which incidentally are the ones the hunters most frequently catch) go to the mouths of the more vulnerable members of the band. Although the human mind is inventive enough to create cultural elaborations which relate only to the creative free spirit of a cultural animal and not at all to the world of practical affairs, it is often rewarding to search for *biological good sense* beneath *social rules*. The universal prohibition of incest among hunting and gathering people, and in virtually every other form of society too, for instance, is surely an example of a biological imperative packaged in a social custom.

Another virtual universal among hunter-gatherers is the allocation of the meat quest to the men while the responsibility for gathering vegetable foods falls to the women: the *collection* of food, unlike its *preparation*, is strikingly non-egalitarian. Anyone who has experienced life in a hunter-gatherer camp knows that meat is much less of a simple body fuel than is plant food. When a man brings meat, whether it is a small spring-hare slung casually over a shoulder, or a large meaty limb sliced from a big antelope, ripples of excitement rapidly spread through the whole camp: the bigger the prize, the greater the excitement.

Not only do hunter-gatherers claim that they prefer the taste of flesh to that of plant foods, but meat also forms the focus of a social interaction that is vital to their way of life: if there is sufficient meat it is carefully shared by the hunter among all the people in the band. Not that one person divides the prey, into, say, 25 pieces and then hands these out to each of the 25 people in the band. Instead, the person who killed the beast gives large portions to his closest relatives and to others to whom he may have a special obligation of some sort. The people who receive something from the first share-out then give portions of their gift to their closest relatives and to those to whom they owe a debt. And so it goes on: the meat is shared in waves along lines of kinship and obligation.

By contrast, when a woman brings home her collection of plant foods she is providing for her immediate family only: plants are generally not shared; they are not an important currency of exchange. Meat undoubtedly is special, not just to eat, but as a crucial part of social intercourse and 'political' structure. A successful hunter can therefore accrue great prestige, not necessarily because of his skill and courage in tracking and killing game, but rather through the prerogative of sharing his spoils he accumulates obligation and respect, the only form of power that prevails in most hunting and gathering communities. Because of his status and ability to provide much-valued meat, an accomplished hunter often has more than one wife.

Although monogamy is not uncommon in practice among hunter-gatherers, it is only rarely a social rule as it is in many western countries. Whether a man has one wife or several depends on his hunting and 'political' skills: the more skills he has, the more women he apparently can claim. Paradoxically, it benefits a man greatly to have several wives because, almost invariably among hunter-gatherers, it is the women who pro-vide most of a family's food: Richard Lee calculates that in the !Kung diet meat accounts for a little over 30 per cent of the food, the rest, of course, being plant materials that are collected by the women. Meat is scarce and difficult to come by, whereas plants are plentiful and do not require tracking, stalking, and killing. In terms of economics, hunting is a high-risk, low-return activity (hunters return empty-handed on four out of five excursions), whereas collecting plant foods is a low-risk, high-return pursuit. Perhaps it is part of human nature that we value those things that are difficult to obtain, while we give scant thought to dependable resources on which our lives depend. Certainly the cave artists of stone-age Europe and Africa were much more obsessed with depicting the vibrant animal spirit of food on the hoof than in capturing the tranquil-lity of whatever was their equivalent of mongongo nut trees.

Once again, one might suspect that the !Kung are unusual in that they can feast all day long on roasted mongongo nuts, and thus have no incentive to hunt for more meat. But, once again, the answer is that the !Kung way of life is very representative of the hunting and gathering existence as a whole. Except for hunter-gatherers who live in the most northerly latitudes where the amount of edible plant material is very low and highly seasonal, all such people eat more vegetable foods than meat: food from the hoof usually makes up somewhere between 20 and 45 per cent of the diet, with an average of 35 per cent (which corresponds almost exactly to the !Kung's).

It would therefore be more accurate to refer to these people,

89

not as hunter-gatherers, but as gatherer-hunters. Only relatively recently have anthropologists ceased to be besotted with hunting as 'the primitive way of life', allowing more recently still the addition of 'gathering' to the description of such people's economies. In spite of its being easier to articulate, we will from now on replace the term hunter-gatherer with the more accurate name of gatherer-hunter. Inertia, tinged perhaps with a little male chauvinism, may well protect the old in-accurate name for some thing to come. However the important point is that in human history the way of life of our primitive ancestors was almost certainly one of gathering and hunting, and not one dominated by a thirst for blood.

When a woman sets off to collect food, usually in the com-pany of several others, she carries two essential items with her, and often a third: she needs a digging stick to excavate nutri-tious tubers; and she must have something in which to carry the food back to the camp. The third item she may have is her infant, who may be anything up to four years old. !Kung women make a large 'bag', known as a kaross, from the hide of a large antelope, and they sling this around their shoulders. The kaross has a dual purpose: for carrying food, and baby too.

The technology of food gathering is therefore very simple: at the least it requires only a receptacle in which to transport the food; and if roots, tubers, and rhizomes are to be on the menu, then one needs a sharpened stick with which to unearth them. During their food-gathering excursions, the !Kung women usually resharpen their digging sticks by a few deft strokes with a crude stone flake. The material aspects of this part of the gatherer-hunting mixed economy could hardly be more basic or less impressive. And yet it is all that is needed. The real skill of food gathering is knowing where to go and when to go there: with a wide range of fruits, nuts, roots, and shoots coming into season at different times of the year and in different places, food gatherers must balance up the proba-bility of success in travelling, say, three miles in one direction to a potentially good source of food, against going four miles in the opposite direction to an even richer source, but perhaps with a lower probability that it is ready to collect just yet.

To make a success of a food-gathering economy you need highly efficient mental maps, not just of space but of time too: you have to know where to go and at what time so that, in terms of economics, you can maximise the return on your efforts. So the key to this type of economy lies in the information and analytical skills inside the head, rather than in fancy technology wielded in the hand.

Although the technological demands of hunting are greater – though only marginally so – than for food-gathering, success once again depends on knowledge and the ability to be able to use that knowledge: hunters deploy more time, effort, and skill in tracking and stalking their prey than they do in dispatching it. They prefer to exercise their wit in getting 10 yards closer to an unsuspecting animal than in trying to improve their technology to produce, say, a flightier arrow.

The simplicity of the technology that supports the highly complex gathering and hunting economy is very important in our concept of our early ancestors. It is very easy, for instance, to make the mistake of judging the complexity of men's minds on the basis of the sophistication of the technology they create. And when looking through the archaeological record it is tempting to attribute an uncomplicated lifestyle and a simple mind to a creature that apparently went about its business wielding uncomplicated tools, as the early hominids did. As the experience of contemporary gatherer-hunters tells us, one could hardly be more wrong.

Gatherer-hunters are very much *a part of* their environment rather than being *apart from* it. And while the women have a keen sense of where and when to find nutritious plants, so the men become accomplished naturalists, eventually gaining an encyclopaedic knowledge of animal behaviour. As part of the research project on the !Kung, Nicholas Blurton Jones and Melvin Konner conducted a series of 'seminars' on animal behaviour with the men in a number of villages. At the end of the day, the time when people sit around the camp fires and chat about the day's events, plan activities for tomorrow, or simply tell stories, Blurton Jones and Konner would encourage some of the men to tell them about the animal world as they saw it. The task was far from difficult as the men evidently enjoyed the whole exercise greatly.

Just like so-called animal lovers in the west, the !Kung were very enthusiastic in retailing stories about 'their' animals. But, unlike westerners, they stuck to the facts. They didn't invent activities where they had not seen them in real life. The truth was clearly fascinating enough in itself. The men knew in detail the important physical features of scores of animals, and they were very familiar with many of the subtleties of the animals' behaviour too.

One would expect the processes of evolution to equip hunters with a curiosity about animal behaviour sufficient for them to be able to follow and kill their prey. But, as it turns out, evolution seems to have over-endowed the !Kung with curiosity: they appear to know more than they need to, and their interest in animals certainly outstrips the necessary minimum. Indeed, in some cases pure interest in the animals themselves actually hinders the hunter's objective of catching a meal: during one 'seminar' a man described in delicate detail the courtship he had seen between a pair of gemsbok, adding that he had been so intent on watching the animals that he forgot about shooting them; the gemsbok escaped before the hunter's mind returned to the more practical matter of getting a meal!

Many times during the conversations Blurton Jones and Konner simply disbelieved things they were told, only to discover later that the !Kung men had been quite correct. As naturalists, the !Kung are at least as accomplished as any in the consciously intellectual western world: they are at least equal to the observations of Gilbert White and Aristotle. But their *explanation* of animal behaviour is poor: they are not greatly interested in the question *why?* When Blurton Jones and Konner pushed for explanations, the men usually offered anthropomorphic interpretations. For instance, they suggested that the reason lions frequently prey on wildebeest is because the meat tastes good to them. And, interestingly, they attribute the motivation of 'withholding' (something unheard of in !Kung society) to lions when they bury the intestines of their prey, and to leopards when they take their hunting prizes up a tree: the lions are 'withholding' the intestines from vultures; and the leopards are 'withholding' the meat from brown hyaenas.

The wealth of information about animal behaviour which the !Kung have, together with the almost total absence of rigorous explanation, contrasts interestingly with the state of ethology in the west 60 years ago when there was virtually no real information about *what* animals did, yet there was a superfluity of theories about *why* they did it!

When the !Kung hunt, something they usually do in pairs, they are faced with solving a constantly changing problem: where is the animal now? which way is it going, and how fast?

is it likely to alter its direction dramatically? how badly is it wounded? how long can it continue before it collapses? These are the kind of questions that are asked repeatedly in a constant computation of the chances of the hunting excursion. Unlike lions and other large carnivores that stalk their prey, pounce, and then kill it, giving up if the animal flees, the !Kung must stalk their potential meal *after* shooting it as well as *before* because the poison on the arrow usually takes some time to bring the animal down.

The hunter feeds a large amount of information into a complex integration and analysis: he has to take account of the season, the time of day, how hot it is, where the wind is blowing from and how strong it is, the nature of the land over which the wounded animal is moving, the tracks of the animal's feet, the condition of the faeces, how much blood is dripping from the wound, where the wound is likely to be on the body, and the displacement of grass, twigs, and bushes in the animal's path. Combine this with a keen knowledge of the animal's likely behaviour under such circumstances, and the hunter has a very good assessment of whether or not there will be meat for supper.

As an example of the kind of clues hunters use to decide whether or not they should persist in their pursuit, Konner tells of an occasion when he was with one of the !Kung who was hunting kudu. That particular hunt was unproductive, so they set off back for camp. On the way they came across a gemsbok spoor which the !Kung thought had been made that morning. They followed the tracks for almost 20 minutes, when suddenly the !Kung abandoned the pursuit, saying that the animal had in fact passed by the previous night, not in the morning. Baffled, and unable to see how the hunter could possibly have come to such a conclusion, Konner asked for an explanation: the !Kung pointed to a mouse track superimposed on one of the gemsbok spoors, saying that as the mouse is a nocturnal animal it could have made its mark on top of the gemsbok spoor only if the large animal had made its tracks some time before the previous night. Simple!

Compared with most hunters, humans are embarrassingly ill-equipped in their olfactory apparatus and are therefore forced to rely on their wit in order to achieve what a lion does merely by sniffing the air. In common with lions, however, human hunters on the track of prey take exaggerated care to avoid making a noise: they move through the bush with a confident quietness born of a true intimacy with the natural world, and they communicate their observations, ideas, and inferences to each other by means of silent hand gestures. How different this is from the ripple of relaxed chatter that runs constantly through the groups of women as they go about their unheralded yet vital task of gathering fruits and roots.

One aspect of the life of most gatherer-hunters that people used to the ways of technologically advanced economic systems find difficult to comprehend is the absence of saving, of investment for the future: gatherer-hunters eat what food they collect, as if there were no tomorrow! True, when a !Kung woman returns with her kaross bulging with both infant and food she has enough to sustain the family for several days. But this is not a matter of deliberate investment in future material security; it is simply a consequence of the efficiency of collecting particularly nutritious foodstuffs.

There is no doubt that the !Kung, and the majority of other gatherer-hunters too for that matter, could lay in food stores if they felt so inclined: they could smoke meat and preserve nuts and other suitable vegetable foods to a much greater extent than they do. But they choose not to. Are they profligate fools with minds untutored in the ways of thrift? This almost certainly was the view of Baldwin Spencer when in 1899 he wrote of the Australian aborigines that they behaved as if they had 'Not the slightest thought of, or care for, what the morrow may bring forth'.

An alternative explanation for this heretical, unwestern behaviour is that they have a firm security about their way of life. As Rodney Needham has said, their behaviour suggests that gatherer-hunters have 'a confidence in the capacity of the environment to support them, and in their own ability to extract their livelihood from it'. Life for most gatherer-hunters is a steady rhythm of work, leisure, and socialising. They move with an easy nonchalance from camp to camp, not with a fatalistic resignation, but with a confident assurance that stems from a true intimacy with nature. How different life is for people who are tied down by a slowly ripening harvest, with the constant nagging worry that one freak storm or pest infestation could wipe out much of a year's food supply.

One feature of the gathering and hunting way of life that impresses economists – and one that leaves biologists more than

94

a little baffled – is that the communities maintain themselves at a level substantially below what the environment could theoretically support. A margin of resource under-utilisation is what an economist might call it. But a more down to earth interpretation is that gatherer-hunters don't push their luck too far: if each year they took 100 per cent of what the environment had to offer, then in poor years there inevitably would be severe deprivation and suffering.

Now, we are not suggesting that from time to time an economics committee of the !Kung – and all the other gatherer-hunters too – comes together to decide what degree of resource under-utilisation they should continue to operate. No. It is a remarkable fact that the gathering and hunting way of life has built into it a margin of safety, a natural buffer against potentially difficult years. To the gatherer-hunters themselves it is certainly an unconscious element of their lives. And for biologists there is a real problem in imagining what kind of mechanism can relate communities to their environment so that they can run below capacity. Nevertheless, there is such a mechanism embedded in the brains of *Homo sapiens*, and very probably it has been there for a very long time, perhaps ever since the gathering and hunting economy first developed several million years ago.

Inextricably linked with the gatherer-hunters' apparent reluctance to overburden their environment, is the frequency with which the women have babies: in the !Kung the average is about once every four years, so that by the time a woman's reproductive life is over she has usually produced four or five infants. As only about half the infants survive to adulthood the woman supplies just enough babies to keep the population stable, with perhaps a very limited growth.

Where does the magic figure of every four years come from? The most straightforward answer is that it relates to the business of rearing the previous child: infants feed from their mothers for at least two and a half years, and often much longer, so that competition from another hungry mouth would be less than welcome; mothers wean their children so late partly because there often isn't enough soft food available for feeding to the soft gums of tiny babies, but also partly because it is all tied in with ensuring a long interval between children. The argument, inescapable, is circular.

Whenever they go out to collect food, the !Kung women take

their young baby with them on their backs, and this is a considerable burden when the child is nearing the age of four. It is therefore simply not practicable for a woman to have more than one child bundled up with the roots and shoots in her kaross: given that !Kung women usually walk almost 3000 miles every year, both on food-gathering trips and moving from one camp to the next, the strain of carrying more than one infant would be intolerable. By some simple mathematics it is possible to compare the degrees of burden a woman would have to carry with birth intervals of two, three, four, and five years. The work that would be involved with the short intervals is gargantuan, but as soon as the interval reaches four years, or just over, the work load drops dramatically and it doesn't decrease significantly more if babies are born even more infrequently. Mathematically, a four-year interval seems sensible. The !Kung, and most other gatherer-hunters too, reach the same figure without the benefit of mathematics!

Because gatherer-hunters the world over come to the same solution, even though they live under a bewildering variety of environmental conditions, one is forced to admit that there is something very basic in the four-year interval. It is certainly the means by which minimum population growth is achieved, but that still doesn't tell us why minimum population growth should be 'thought', biologically, to be a good thing among gatherer-hunters.

Although prolonged breast-feeding does suppress ovulation to some extent, it is by no means a foolproof method of contraception, as many people discover to their cost! For this reason the social rules of many gatherer-hunter communities often demand a long period of sexual abstinence after a baby is born; abstinence rarely stretches beyond a year, however, as this would seem to stretch self control just a little too far. It comes as no surprise, therefore, that abortion and infanticide (especially killing girl babies) are common, though very much covert, parts of gatherer-hunter life. For instance, when Chagnon arrived at his Yanomamo village a woman, Bahimi, was pregnant. When the baby – a boy – arrived, however, Bahimi promptly killed him, explaining tearfully to Chagnon that he would have taken milk away from her other infant, Ariwari, who was still nursing at the age of almost three. Rather than risk Ariwari not surviving the trauma and hazard of early weaning, Bahimi killed her new-born son. Sad, but necessary.

It may wound the sensitivities of people brought up in modern western culture in which life is sacrosanct at any cost – and often the cost can be very great – but we have to conclude that systematic infanticide must have been very common in the long human career. The mobility of nomadic gatherer-hunter way of life demanded it. Only when relatively sedentary life became possible with the invention of agriculture did it make sense to reduce the effective interval between children. And with that came the first rumblings of the population explosion.

The gathering-and-hunting existence demands that people should be relatively thin on the ground: an average of one person in every square mile of land is normal for gatherer-hunters. Not that these people dislike being close to each other, but the economics of meat and vegetable collecting mean that a large area is necessary in which to operate them. And we also begin to look at the psychology of gatherer-hunters and the way that this may have changed when people took up hunting. For instance, is the Yanomamo's fierceness related to their need to protect carefully cultivated gardens? How much does the relative mobility and territorial freedom of gatherer-hunters and the relative immobility of agriculturalists influence the psychological attitude to daily life?

An individual !Kung camp is a striking social contract with the ecological imperative for dispersal: the people live, work, and play very close together, often in intimate and literal contact with each other. Each family has a small shelter which, in the wet season, is made from long poles bent over to form a dome-like structure which is then covered with grass and leaves. The inevitable hearth is the focus of family life where food is prepared and eaten. The huts are usually arranged in a rough kind of circle, forming a communal central area which is the scene of dancing and the first sharing of meat from a large animal. There are no separate living areas and 'workshop' areas: if, for instance, a number of men want to make poisoned arrowheads they usually congregate around someone's hearth where they exchange stories as they work. Everyone's hearth can be the centre of a specific activity at some time during a camp: there are virtually no specialists in gatherer-hunter communities.

The arrangement of the !Kung camp can give us inspiration when we are peering back through the archaeological record. If ancient camp sites contain some indication of a collection of

97

hearths arranged in some form of regular pattern they might be giving us the first glimmerings of the emergence of the family within a social group, something that is unique in the animal world.

Although the exact number of people in !Kung camps varies from site to site, the average is around 25. And 25 turns out to be another significant number for gatherer-hunters, for it is the average band size for the vast majority of such peoples. Presumably 25 is an optimum number of individuals for operating the unique mixed economy of gathering and hunting, both for exploiting an area of workable size, and for cooperating with each other to achieve that exploitation. As each person appears to require about one square mile of land to ensure an adequate supply of food, a band of 25 people needs 25 square miles. Suppose, now, that camps had 100 members, then it means that they would have to exploit the resources of 100 square miles: to do that would demand that they were an unusually mobile group. The 25-member band makes sound biological sense, as one might expect.

Most !Kung men hunt in pairs, an arrangement that makes minimum demands on cooperation between them. However, whenever someone is lucky and brings down a large animal all the men cooperate in butchering the beast and in carrying the meat back to the camp. By contrast, people such as the Bihor in India and the Mbuti in the Congo basin often hunt animals by driving them into long nets which are usually strung in a long line through the trees; this type of hunting is impossible unless individuals work closely together. And a currently popular view of human evolution points to such cooperation between large groups of hunters as a key element in the emergence of human characteristics.

Certainly, there must have been many occasions in the past when brief opportunities of a hunting bonanza encouraged people to come together in large groups in order to exploit it: the archaeological record has evidence of this type of activity, with the half-a-million-year-old remains of a slaughtered troop of baboons at Olorgesailie in Kenya, and large numbers of dead elephant of a similar age at Torralba in Spain. Surely the real focus of cooperation in emerging human society, however, was in operating the mixed economy as such. With a division of responsibilities between people, and with each person depending on every other for survival, then an ability and a motivation to work towards the same objective would have been an essential evolutionary imperative. It would simply be impossible to exploit successfully the wide range of resources available to gatherer-hunters unless they cooperated closely together. Cooperation must be a very basic motivation in human nature.

Although the existence of true human families (males, females, and offspring) within bands is unique in the animal world, bands as such are not: chimpanzees, baboons, and gorillas live in troops, some of which are similar in size to gatherer-hunter bands. But unlike the higher primates, the human's bands are part of a larger structure – the tribe, an organisation for which there is no equivalent among other animals. The tribe, or rather the dialectical tribe as anthropologists frequently refer to it, gives us a third 'magic' number for the gatherer-hunter way of life: the first was the four year interval between children; the second was the size of the local band – 25 individuals; and the last, the size of the dialectical tribe is 500.

In spite of some marked variations, which can be explained by local ecological conditions, the size of dialectical tribes in many different parts of the world homes in unerringly at around 500. Tribes are not just inventions of anthropologists anxious to analyse the lives of technologically primitive people by carving up their social structure into objective but somewhat artificial units. The people themselves are very much aware of dialectical tribes, and they frequently identify very strongly with them: the overt signals of group identification are particular styles of dress or body decoration, or idiosyncratic ways of making otherwise utilitarian objects – but deepest of all is a characteristic language or dialect, hence the term dialectical tribe.

Why should tribes exist? Are they biological necessities or cultural creations? Although the expression of tribal identity is through the medium of cultural invention, it seems that their role is in constituting a population large enough to have roughly equal numbers of males and females. Families are the fundamental units of economic activity; bands are the basic conglomerates of the gatherer-hunter economy; and tribes are the smallest breeding populations within which bands can operate.

If you were a young, virile male coming up to marriageable

age in a gathering and hunting band of about 25 people, your chances of finding even one female of the appropriate age would be pretty slim; and the probability of exercising any choice or whim would be perilously close to zero. What is more, even if there were any potential mates over which to cast a discerning eye, they would almost certainly be closely related to you. And just as the animal world appears to organise its affairs so as to avoid incest as much as possible, so too do humans. The only choice, then, is to look elsewhere for a mate.

In the best tradition of human biology, the driving force of necessity is frequently accompanied by opportunity for incidental advantage: through having to find mates in neighbouring bands rather than inbreeding within one's own band (if there happened to be enough potential mates available that is) gatherer-hunters automatically weave a kinship network over a wide area. That network usually spreads until the numbers of people in the separate bands total 500.

Sherwood Washburn, an American anthropologist, once decided to sit down and work the whole system out from first principles: given the average size of gatherer-hunter bands, the average birth rate and child mortality, the usual birth imbalance between infant boys and girls, and given the need for avoiding incest, how small a population would be sufficient to ensure a roughly equal proportion of males and females of marriageable age? The answer, of course, comes out at 500. Once again gatherer-hunters reached the correct solution without recourse to a computer: through the ages the cutting edge of evolutionary pressure has carved out the most efficient system.

Biological necessity has, of course, been dressed in cultural garments so that most gatherer-hunter societies insist on marriages between bands rather than within them: it is called exogamy (by anthropologists that is). Exogamy is the fountainhead of band interaction in the world of gatherer-hunters, and as such it injects an acute awareness of kinship into these people so that they know very clearly to whom they are related and with whom they may become mates. Generally it is the rules of marriage that make exogamy essential if people are to find mates. For instance, parallel cousins (father's brother's children and mother's sister's children) are usually forbidden to marry, whereas cross-cousin marriages (father's sister's children and mother's brother's children) are often preferred. Sometimes it becomes extremely complex so that a man may have to find a girl who is at once his mother's brother's daughter, his mother's mother's brother's daughter's daughter, and his father's sister's daughter's daughter. It *is* possible!

Baboons and chimpanzees indulge in exogamy too: young sub-adult male baboons usually leave the troop in which they were born and join a new one; by contrast, in chimps it is usually the young females who take the initiative and go. Clearly there is a biological imperative driving these animals that effectively achieves exogamy – but it would be surprising if they have rules about it as well. The interesting point about humans is that we have generated social rules – some of which are unbelievably complicated – that at once prevent incest and ensure important interactions within an effective breeding population. And it perhaps should not be surprising that in many cases the rules have extended the system far beyond the basic biological advantages. In many cases women have become objects of exchange for the purpose of smoothing the path of trade or strengthening alliances in the event of hostility with other parties.

The Yanomono, whose culture drives them relentlessly into repeated combat with each other, use women quite openly to make allies by having them marry into villages that would be suitable fighting partners. And, although each village is materially self-sufficient and therefore has no reason to trade with others, they do trade with each other, but this too is a means of strengthening alliances. The Yanomono of course have something to fight about: they have crops to protect!

Pure gatherer-hunters the world over have so many points in common in their way of life that we may be bold enough to suggest that at least some of them stretch back into the long career of this mixed economy. Bands are almost certainly ancient organisations; indeed there is some evidence that even some of the early Pleistocene camp sites (about two million years ago) were occupied by *about* 25 individuals – it is impossible to be sure of course. Dialectical tribes may well have long been a part of human history; one can even argue that the need for such structures and the concomitant requirement for handling kinship terms were important engines in advancing the evolution of spoken langauge. But what we can be certain of is that without a keenly developed sense of cooperation and a facility for sharing, the gathering and hunting way of life would never have become as successful as it undoubtedly was:

its long history and its ready adaptation to every conceivable ecological niche the globe has to offer testifies cogently to that success.

How did the gathering and hunting way of life develop in the first place, in our Pliocene past? And *what* made such a daring departure from normal primate ecology so successful? We ask these questions in the next part of our story.

CHAPTER 7

Carrier bags, altruism, and the first affluent society

Midway through 1973, on the low, green hills just south of Lake Elementeita (one of the many soda lakes in Kenya's Rift Valley), a large male baboon scattered a herd of Thomson's gazelle – mainly mothers with their offspring – by rushing at them. The antelopes fled in all directions, the mothers desperately trying to keep their young by them. The baboon's intent was unmistakable: he wanted a Tommy for his supper. Eventually he got one, or rather a share of one, but not until after a remarkable series of events that were to make the baboon and his troop very special.

American primatologists Robert Harding and Shirley Strum had been observing this troop, which they'd called the Pumphouse gang, for many years in order to learn more about the social organisation of open-country baboons, animals that occupy a habitat that is probably very similar to places where some of our early ancestors lived. When Strum saw the male – Rad – make his dash at the Tommies on that day in 1973 she knew that this wasn't the first time he'd set his mind on a young Tommy for a meal. More than any other individual in the Pumphouse gang, Rad had a taste for meat, and his hunting forays were often successful. On this occasion, however, Rad mistimed his pursuit and had to give up in order to cool off (baboons are not really designed for long, hot chases).

As luck would have it, the fleeing Tommy was making his escape, unwittingly, in the direction of another adult male baboon, called Sumner, who happened to be nearby. Sumner snatched his opportunity and tried to capture the frightened prey, but the antelope was too quick for Sumner and it escaped his clutches too. With Rad resting and Sumner on his ill-fated pursuit, two other Pumphouse males – Big Sam and Brutus – arrived on the scene, and something like a relay began. The young Tommy's life ended when Big Sam chased the terrified antelope into the arms of Brutus.

Without doubt this relay chase was an accident, a combination of fortuitous timing. But, as proto-cultural animals, baboons are quick to learn tricks that might benefit them, and although most of the Pumphouse gang's hunting thereafter remained more or less solitary affairs, there were numerous occasions on which a group of males combined in successful predatory relays. That baboons should hunt and kill animals as fleet of foot as Thomson's gazelle is interesting enough, given the long-held belief that they are principally vegetarian creatures. But that they should occasionally indulge in an activity that brushes very close to cooperative hunting is fascinating. The world's primatologists were intrigued.

Not that baboons were the first large primates to display hunting skills: for a number of years observers at the Gombe Stream reserve on the shores of Lake Tanganyika had seen chimpanzees stalk, chase, and kill several different types of victims, including colobus monkeys and young baboons. And occasionally too the Gombe chimps appeared to gang up on their chosen prey and cooperate in its dispatch. The discovery that the chimps' lesser evolutionary cousin, the baboon, also hunted in more than a desultory way caused people to ponder on what it might tell us about the emergence of primitive humans from an ape-like stock. Are we seeing here in these modern primates the seeds of a predatory lifestyle that, in our primitive ancestors, grew and blossomed into a hunting way of life? Or is the occasional excursion into meat-eating just one facet of an opportunistic subsistence shared by baboons and chimps?

These were the kinds of questions that were on the lips of many people enquiring into the origins of humanity. And it was against this background that Robert Ardrey published in 1976 his 'Personal conclusion concerning the evolutionary nature of man': he called his book *The Hunting Hypothesis*, and the thesis is simple: that it was only when males of our ape-like ancestor seriously took up hunting that we began to accelerate down the road to humankind. He says that 'Man is man, and not a chimpanzee, because for millions upon millions of years we alone killed for a living'.

Ardrey is not the only proponent of hunting as the key behaviour that made us what we are: the idea is very popular, and indeed meat-eating is incontrovertibly a very special element in the human career. Proponents suggest that it was the intellectual skills demanded by organised hunting, and the high levels of cooperation between hunters that go along with it, that propelled the human brain towards its unparalleled evolutionary position.

We *are* the only primate who includes a substantial amount of meat in our diet: compared with chimpanzees and baboons, for both of whom meat makes up about one twentieth of their menu, most non-agricultural, technologically primitive people have diets composed of at least one-fifth meat and four-fifths vegetables. And in the archaeological record there *are* petrified glimpses of organised meat eating stretching back at least two million years, and probably closer to three: the simplest explanation of sites, such as the KBS camp on the shores of a now diminished Lake Turkana, in which stone tools and animal bones are found together is that the hominids collected the bones while they were still encased in red meat and took them to a suitable spot to be part of a meal. But it is a giant leap to go from these observations to say that organised systematic hunting was the *primary* force in the birth of humanity.

Perhaps in response to the male chauvinism implied in the hunting hypothesis, a number of people have proposed its direct counterpart, the gathering hypothesis. For instance, American anthropologists Nancy Tanner and Adrienne Zihlman see the first glimmerings of humanity beginning to shine when a group of female hominids went off into the bush to collect and bring back food for themselves and for their offspring.

Taking the lives of virtually all higher primates as a model for those of our ancestors, Tanner and Zihlman focus on the female as the pivot of all social interactions: the bond between a mother and her infant is the strongest a primate ever experiences; and the mother is the principal educator, both in practical affairs and in their involved social life. The males, in a sense, are necessary only because nature has chosen to make sex an inescapable part of reproduction!

Compared with males, then, females have a tremendous investment in the offspring (males have an investment too, of course, their genes, but if the womenfolk are prepared to nurture the products of those genes, the offspring, the males need not exert themselves; here we find ourselves in the deep and muddy waters of sophisticated genetics, so we will avoid this particular issue for a while). The females' pivotal position in the proto-hominid social group, Tanner and Zihlman argue, put into their hands the power to exploit technological innovation. The first tools, they suggest, were invented '*not for hunting* large, swiftly moving, dangerous animals, *but for gathering* plants, eggs, honey, termites, ants, and probably small burrowing animals' (their italics).

In this view of the early hominid world there were advantages to be had in having a menu not unlike a modern chimpanzee's, but which was eaten not during a daily round of social foraging but after it was collected and taken to some form of camp where mothers and offspring gather. Meat became important only after this initial shift in eating habits, thus including males more closely in the social fabric. The major strength of the gathering hypothesis is that humans are the only primates who *collect* food to be eaten later; they are *not* the only primates who hunt. On their occasional hunting forays baboons and chimps may well be re-enacting scenes very similar to those played out by hominids between four and 15 million years ago. But no primate gathers food – ever!

So, on the one hand we have the hunting hypothesis, and on the other, the gathering hypothesis. One argument is male-centred and depends on an early commitment to a quest for flesh. The other focuses on females and sees improved child care as an evolutionary cutting edge for advancement. Both have arguments in their favour. Both can be criticised on specific grounds. But more important, both miss the essential element of the uniquely human way of life: the economic pact between suppliers of meat and suppliers of vegetables. The heart of this economy is the sharing of two different types of food: it was *the first mixed economy*. The question we have to answer is why it evolved at all?

New styles of behaviour, and their associated anatomical expressions, emerge from the evolutionary melting pot, not through the whim of some quixotic guiding hand, but because they are in some ways advantageous. And this rule must apply to our early *Homo* ancestor's dual approach to subsistence just as much as it does to the elephant's trunk, the zebra's stripes, and the termite's complex chemical signalling system for controlling colonial life. By adopting an economic approach to life in place of a literal hand-to-mouth subsistence, we must infer that these hominids were better able to exploit the environment in which they lived, better that is than their australopithecine cousins who eventually failed the evolutionary challenge and slid into oblivion around one million years ago. How? Specifically, through the separately organised collection of animal and plant foods our ancestors won for themselves a bigger share of the energy resources available to them: and much of biology is simply to do with the best way of getting energy.

Many people like to compare the human diet with that of pigs and sometimes bears too, saying that it is omnivorous. In a sense this is true: as opportunistic creatures our ancestors would have rarely passed up the chance of an edible morsel, whatever it was. Opportunism was a key feature in our evolutionary success, and it is a quality shared to varying degrees by many higher primates. For many years virtually all higher primates were classified as basic vegetarians, as fruit-eaters or leaf-eaters, for no better reason than no-one had really bothered to find out the truth. If you were a chimpanzee you would not spend your life gormandising on bananas and figs: you'd have a rich and varied mixture of fruits and berries, leaves and shoots, reed and wood pith, nuts and seeds, resin and bark, buds and blossoms, eggs, grubs and insects, honey, fledglings, lizards, and several types of mammal, including infant and juvenile baboons, young bush-pigs and bushbucks, colobus monkeys, and red-tail and blue monkeys, and more besides.

The chimpanzee is truly omnivorous, and most baboons are not so very different, except that they eat less fruit and more seeds and roots and other tough small items; the meat on the menu of the olive baboons near Lake Elementeita is mainly young Tommies and spring-hares.

For chimps and baboons there are no such things as 'meal-times': they spend much of their day in desultory feeding, interspersed with the serious business of a complex social life, involving grooming, supportive interaction with relatives, flexing a little social muscle with competitors, and keeping a sharp awareness for a chance to mate. Though in their different ways both chimps and baboons are intensely social animals, their feeding is an essentially solitary business: they forage together in groups small and large, but each individual plucks and eats its own meal – they do not share. If modern humans were to follow this pattern, people might go along to a dinner party each equipped with their own food which they'd cook and eat separately, all the time gossiping between each other about the things that people normally gossip about on such occasions! It is solitary feeding in a social context.

But humans don't behave like this: we share our food, and our argument is that we've shared it for many millions of years. Sharing, not hunting or gathering as such, is what made us human.

If we try to look back to some unspecified time at a formative period in the early human career, we have to ask more specifically how sharing might have been an attractive prospect in biological terms. For a start, evolutionary advancement rarely occurs in the midst of a population whose needs and wants are fully provided for. Natural selection has to have some kind of cutting edge, and that cutting edge is usually found slicing away at the fringes of a population where it is a little harder to make a living and where the dangers are a shade greater than in the biologically untroubled middle.

If, say, six million years ago, our ancestors could sit back all day and gorge themselves sick on nuts and berries and succulent roots, then we would not be here today pondering on our origins. Some members of the human family may well have been sufficiently lavishly provided for them to lead such a primitive sybaritic life, but undoubtedly many were not. And it was the have-nots in whose interests it would have been to develop a new way of life. (Once again, we do not suggest that any creature actively decides its evolutionary future: in any population of animals there is a spectrum of behaviours and body shape, and it is the ones that are most suited to the prevailing environment at any particular time that survive best and thrive in the future.)

Throughout the whole of humanity's long evolutionary career plant foods have been the primary food: the large size of our cheek teeth and their unusually thick enamel tell us that, and this applies from *Ramapithecus* right through to modern *Homo sapiens*. We know too that through that period meat gradually became a more and more important item on the prehistoric menu for some of the hominids, but, except for people in unusual situations such as the Copper Eskimos who eat nothing but meat in their frozen homelands, flesh rarely rivalled plant foods as the staple food: the steady reduction in the size of the cheek teeth from the relatively massive molars of *Ramapithecus* to the still generously endowed, though less formidable, jaws of modern humans points to a change in our ancestors' diet; and so too does the increasing frequency with which one finds animal bones associated with stone artefacts as one scans through the archaeological record towards modern times.

Before meat began to take its place in the unique mixed economy of early *Homo*, we can guess that our hominid ancestors tasted flesh every once in a while, just as chimps and baboons do today. We don't know what else they were eating, though it probably wasn't identical to modern chimp or baboon menus: if the ancient diet had been like these modern animals', then we'd expect their teeth to be more alike than they are. Ancient *Ramapithecus* may well have depended heavily on grass or roots or seeds, diets that would have demanded their efficient grinding equipment, but it would be surprising if those petrified pieces of jaws now lying in museums in many countries of the world did not occasionally savour the taste of meat too. *Ramapithecus* was probably omnivorous to a degree, like its modern cousin the olive baboon.

The development of a hunting-based economy has also been invoked as the cause of another uniquely human characteristic – the ability to walk upright. It is easier to pursue potential prey and wield deadly weapons if you are upright, the argument suggests. But as no one who is absolutely honest with himself can justify his voice in support of any one particular theory of the origins of our somewhat unusual way of getting about, let us just for a moment think of fully bipedal *Ramapithecus* some six or so million years ago who have not yet begun to eat more meat than today's chimps and baboons. Many people may object to this picture, and they may eventually be proved right to do so. But at the moment there is truly no solid evidence one way or the other.

Starting with our heretical premise of early hominids having adopted upright walking for some unknown (and unknowable?) reason, we can set them off in search of hitherto unexploited rich packages of food: meat. But we are not talking about newly feathered fledglings or even a young antelope as potential meals; we are interested in occasional bonanzas – a hippo perhaps, or a hefty water buffalo. The prospect of a group of diminutive hominids – perhaps around four feet tall in their stockinged feet – setting out to bring down one of these bulky and dangerous animals with their bare hands is perhaps not very convincing. But, with a little luck, they would not have had to: enough of these creatures expire naturally or are dispatched by carnivores to provide a relatively rich opportunity for scavenging.

In the summer of 1969 two American anthropologists, George Schaller and Gordon Lowther, put the idea of hominid scavenging to the test in a simple, ingenious, and direct way: they tried it for themselves. They camped for a week in the woodland banks of the Mbalangeti River in the Serengeti, an area frequented by zebra, wildebeest, impala, topi, buffalo, gazelle – and lions. They spent their days walking around looking for young, sick and dead animals which, if they had been searching seriously for food, they could have collected. They found four fresh lion kills, but all that was left of the victims was marrow and the brains, enough to provide good meals, but not exactly a gargantuan feast.

Lions in fact frequently eat just a small amount of their prey, the rest providing many meals for scavengers such as vultures and hyaenas. The lions around the Koobi Fora camp were particularly active during the summer of 1976, feasting off the herds of zebra that abound there. One morning the camp woke to find the sandy shore criss-crossed with lion tracks and churned up by the countless hooves of stampeding zebra. And there, in the middle of it all, was the virtually intact corpse of an adult female zebra. The lion had ripped open the zebra's belly, chewed at its groin, and departed. Had the time been two million years ago, the hominids in the Koobi Fora camp would have woken to find a real meaty bonanza on their doorstep.

As well as lion kills Schaller and Lowther also found a recently expired bull buffalo, an enormous animal that had died of either disease or old age. They estimated that they could have scavenged perhaps 259 kg of meat from the carcass. The 'hominid scavengers' also came across a sick, abandoned zebra foal and a blind giraffe, which between the two would have supplied another 200 kg of meat. This is a very splendid haul for just two people searching during one week. They knew too that in the right season they could be certain of catching very young animals that mothers leave hopefully hidden in the grass: as with the young of many animals, Thomson's gazelle fawns crouch low in the grass and 'freeze' when danger threatens and many predators simply don't see their immobile bodies; the more cunning hominids might have capitalised on this method of defence.

Perhaps Schaller and Lowther were extraordinarily lucky in their 'scavenging'. Perhaps if they'd tried a week earlier or a week later they would have finished up empty handed. This is quite possible, and this is the point about a dual approach to a subsistence economy: when you do come across a dead or dying buffalo, for instance, the reward is enormous. But for the lean times when the hyaenas get there first or potential carcasses are too thin on the ground to be discovered easily, you need some kind of back-up. And this is where plant foods are important. Glynn Isaac, a brilliant archaeologist and long-time member of the Koobi Fora research team, favours scavenging as a route for breaking into the mixed economy, and he describes the reliable plant foods as an insurance policy, a way of feeding adequately between the occasional scavenging bonanzas.

Initially, of course, active scavenging would not have been part of the normal daily round of our early ancestors: they would have taken their opportunities as they came across carcasses during their foraging for nuts, roots, and other plant foods. But they would have soon learned that the wheeling vultures in the sky signalled a potential meal if they could get there first. Eventually the search for meat would have become more than a desultory affair, with some individuals spending much of their time doing it, just as Schaller and Lowther did in the Serengeti. And they would have also learned to look in the trees for their fleshy food as well as on the ground because leopards take their kills up into the branches out of the reach of the ever-hungry hyaenas. Hyaenas can't climb trees, but the hominids certainly could, and if they were willing to take the risk of confronting an angry leopard who returned to find his larder being raided, then they would have eaten well.

Through scavenging meat these hominids would have tapped a greater energy source than is exploited by, say, modern chimpanzees, in spite of their occasional predatory habits. By including substantial amounts of meat in their diet, in addition to fruits, shoots, nuts, berries, etc, our ancestors put themselves in a new league: they were not carnivores in the way of the lions and leopards they stole from; nor were they basic omnivores in the style of chimpanzees and baboons; they were super-omnivores, capable of making a good living in what might otherwise have been a demanding environment. The fact that gatherer-hunters can exploit both plant and animal foods, rather than having to rely on just one or the other, is the prime economic argument against both the hunting hypothesis and the gathering hypothesis. Neither of these latter two life-styles offer as broad a resource base as that enjoyed in a gathering and hunting way of life.

Incidentally, although chimps and baboons give chase to more or less any small animal that comes close and is a potential meal, they *never* scavenge. Primatologists studying these animals have even resorted to putting the bodies of animals that are normally enthusiastically hunted and devoured with relish near to troops of chimps and baboons, but in all cases the corpse is sniffed, prodded – and then ignored. Why, no one knows. If scavenging rather than hunting was the seed of emerging humanity, then clearly these modern primates don't have it in them.

But the really important thing about our ancestral hominids was not *what they ate*, but *how they collected it*. For the most part, the food that finishes up in an individual chimpanzee's stomach has been collected by that same animal. By contrast, our ancestors invented an economy in which some individuals collected mainly one type of food while others collected mainly a different type: there was a division of labour, and the products of separate food quests inevitably implies what anthropologists term 'postponed consumption': instead of eating your meal straight off the tree or out of the ground, you gather up a bunch of roots, berries, and grubs, and take them back to a camp where you meet with other individuals to share and eat the food. Hominids collected food and carried it back to their camp; the only place a chimpanzee collects food is in its stomach! So, although it is true to say that hominids were omnivores in what they ate, it is not a very good description of how they organised their subsistence economy.

One social consequence of the greater efficiency of the gathering and hunting economy in exploiting available energy resources was, almost certainly, more leisure. Meat is an energy-rich source of food, so much so that successful carnivores such as lions and wild dogs spend at least 20 hours a day simply resting. As our ancestors become more able to satisfy their food requirements easily through their dual-approach economy that included meat, they too would have had to devote less of their waking hours to an unrelenting food quest. Perhaps two or three million years ago early *Homo* wasn't as efficient as the modern !Kung or other gatherer-hunters, but it would be surprising if they hadn't enjoyed a good deal of spare time, certainly more so than modern socially oriented chimpanzees. So, in the words of Marshall Sahlins, the first mixed economy brought with it leisure and the *first affluent society*.

Much of what we are saying is, of course, inference: the virtually complete absence of any direct sign of plant foods on camp sites more than a million years old leaves us playing with ideas rather than handling fossil remains as we are able to do with animal bones. But the scenario must be pretty solid because we are dealing with something that is particularly human, something that must have stretched a very long way back in our history, something that, as Ardrey put it, made man man rather than a chimpanzee.

If a division of labour was so important, how was it organised? Again we are guessing, but it is not unreasonable to suggest that, as with virtually all non-agricultural peoples, the meat was the business of the men while the insurance policy was in the hands of the women. Why the division should be so clear-cut in modern human gatherer-hunters is a little puzzling: women with a small infant to care for may be hampered on hunting forays, of course, but there is a niggling intuition of something deeper than simple practical matters. In chimpanzees and baboons too, hunting is an almost exclusively male preserve. Once again perhaps the females' commitment to child care may prohibit them taking off on hunting excursions. The females are undoubtedly the focus of social life in these primates, and they usually have a string of offspring of increasing age somewhere close to them. Those few female chimps and baboons who do show an interest in hunting and meat eating invariably pass on their penchant to their offspring, thus demonstrating the females' key role in education.

Although we reject their hypothesis for the emergence of humanity, Nancy Tanner and Adrienne Zihlman are almost certainly correct when they suggest that the first implements in the hominid tool-kit were simple digging sticks and some kind of container in which to carry nuts and roots. It is easy to carry a large amount of meat: you just sling the animal, or a severed limb, across your shoulder. But a pile of berries presents a technological problem: without an efficient container you either eat them on the spot, or leave them to rot. Chimpanzees strip leaves from twigs to form a probe for catching termites; they also sometimes use a bunch of leaves to soak up water or the juices from the braincase of a slaughtered victim; they even occasionally use a stone to smash open hard nuts. None of these tools actually changes the animals' lifestyles very much. But the invention of a primitive container – the first carrier bag – transformed the early hominids' subsistence ecology into a food-sharing economy. The digging stick may have come before or after the carrier bag, but, important though it was, it lacked the social impact of the container; the digging stick may have made life easier, but it didn't usher in an entirely new lifestyle.

There are, of course, no signs of this major technological revolution in the archaeological record: as the first containers were probably made of woven leaves or bark, they simply vanished without trace. The idea of an ancient woven carrier bag may seem somewhat far-fetched, but in fact chimps and gorillas are adept weavers. These closely related apes both make sleeping nests for themselves by threading branches, twigs, and leaves through each other. Even captive orang-utans, with their less dextrous hands, have been seen to weave strands of straw together, seemingly to alleviate their boredom. Surely, then, it was not beyond the wit of the early hominids to extend this basic ability in order to construct a crude carrier bag.

Because digging sticks are more effective when they are relatively sharp it is a fair guess that the early hominids very soon learned to whittle branches to a point using stone flakes they happened to come across. From there it is but a short intellectual leap to manufacturing a cutting edge by breaking one stone in half by striking it against another, or, more crudely, by smashing it to the ground. The point here though is that the first stone tool technology, if one can glorify such activity by the term, was very probably needed to fashion other tools of wood. Although the product of such activity is hardly impressive, the successive steps of intentionally making a stone cutting edge which is then used to shape a wooden implement is an intellectual order of magnitude removed from a chimp stripping leaves from a twig in order to make a termiting stick.

Such stone tool 'industry' could have gone on for millions of years leaving absolutely no detectable archaeological trace: our ancestors would have struck their stone 'knives' wherever and whenever they wanted them, discarding them immediately. It is an exciting thought that such tools, which might be as old as six million years or even more, are probably scattered throughout ancient deposits in many parts of Africa. But the excitement is tempered by frustration because we would almost certainly be unable to recognise one even if we found it. Only when the making of stone tools became a concentrated affair, both in location as well as in different tool types, is there a good chance of identifying them as the products of protohuman hands.

As we look back through the major archaeological sites in East Africa – Olduvai which is almost two million years old at the base, the KBS site at Lake Turkana with a date slightly more than two million, and so far the oldest of all, a site at the Hadar in Ethiopia which has tools that were made an undisputed two and a half million years ago – there is, as one would expect, less and less sophistication in the tool-kits. But the important point to ponder is that throughout this time hominids were coming together for organised technological activity. Although it is not always easy to say whether such remains represent an ancient camp site, almost certainly many of them do, and are thus a clear indication of a distinctly human form of activity. The camp was the focus of social life and of the gathering and hunting economy. It was here that the food was shared.

At some turning point in our history the primitive *Homo* males began to take a serious interest in hunting as another way of providing meat: operated together, scavenging and hunting would have been more productive. In their first tentative excursions into meat collecting (both dead and alive) the early hominids were, of course, entering the reserve of well-established and efficient carnivores: lions, hyaenas, wild dogs, leopards, sabretooth tigers, all had been in the meat-eating business for a very long time. The competition was potentially very tough. But, as it happens, most of these animals operate either in the cool of early morning or in the gathering darkness of nightfall. The hominids, like most modern higher primates, including humans, are daytime creatures. So, back in the Pliocene there may well have been an opening for a new scavenger/hunter to establish itself in daylight hours. Our early ancestors seemed to have taken that opportunity.

The earliest hunters must have dispatched their prey with the crudest of weaponry: a branch used as a club perhaps, a well-aimed rock, or maybe a digging stick became a spear when the chance of game appeared. Those people who see a weapon in every ancient stone implement must surely be mistaken because the technology of killing, say, an animal mired in a swamp is very simple indeed: there is no need to manufacture specially shaped missiles to hurl at the unfortunate victim: any heavy rock would be effective. And if hominids threw missiles at fleeing prey, the same holds true there too.

The oldest undisputed remains of a spear in the archaeological record were discovered in Clacton, a town on the east coast of England. At a mere quarter of a million years old, this 15-inch point carved from yew is a relative youngster: crude spears may well have been employed with devastating effect two or more million years ago. Certainly, if one ponders on the remarkable ability of modern humans to calculate the required trajectory and force in order to hit an object accurately with any kind of missile, from a stone to a heavy javelin, then it seems indisputable that throwing things with serious intent has long been a human activity. Compared with us, our close relatives the chimp and gorilla are mere amateurs; their intent may often be serious, but their aim is usually poor. There does appear to be one exception to this, however, and that is a gorilla at San Diego Zoo: the magnificent creature repeatedly embarrasses zoo officials by hurling his dung at visitors as they pass by his enclosure in the sightseeing train!

Technology, early in the emerging human career, must therefore have focused on collecting plant foods rather than on the much vaunted hunt. And sharing the spoils of scavenging in the early days may have been possible without tools too, just as modern chimpanzees tear carcasses apart with their hands and teeth. But it must have struck someone pretty soon that it would be a good idea to slice the meat with the same types of stone flakes that were used to sharpen digging sticks. Simple chunks of rock would have been more than adequate to smash open bones to release the juicy marrow. Only when practical affairs became really quite complicated, involving perhaps preparing animal skins, did the tool-kit need to be diverse. But we should always remember that any stone tool-kit has at least three components: implements for preparing meat; implements for preparing plant foods; and implements for making implements, either of wood or bone.

The social life behind this technology is, of course, what we are really interested in, and that, as we have suggested, focuses on a reciprocal food-sharing economy. No other animal shares food in this way. True, some of the social carnivores share their hunting prizes. Perhaps the best example is the African wild dog, a beautiful animal that displays some very human characteristics. For instance, when a pack goes hunting they leave behind in their den the young who are cared for by one or perhaps two adult females, plus any injured or sick member of their group. When the hunters return they regurgitate meat for all who were left behind, including the incapacitated animal. Such care-taking is unusual in the animal kingdom, and it is probably a result of the wild dog being such a highly social animal. The point about their food sharing is that all the animals engaged in the food quest are pursuing the same type of food: there is no dual approach to the subsistence economy.

If we come a little closer biologically to the hominid line, to chimps and baboons, we can see an interesting and instructive difference in these animals' social manners when they have the good fortune to include meat on their menu. In both chimps and baboons it is mainly the males who do the hunting, but there the resemblance ends. Baboon males are impressive creatures, much bigger than their females, and sporting fine sharp canines. There is often a good deal of tension between the males, and usually each individual knows who is boss: there is a pretty well-defined hierarchy. When one of the males catches a young Tommy, for instance, many of the other males are quickly on the spot, attracted by the cries of the victim. From then on tempers are on edge, and frequently the meaty meal finishes up in the hands of the dominant animal. There is no such thing as sharing, except perhaps between a male and a special female with whom he may be consorting, and between a female and, if occasionally she manages to procure some, her offspring.

Chimps are different. Not that a successful hunter sits around handing out his spoils in a free and generous manner, but there is a good deal of sharing. Most people who are lucky enough to witness a successful kill by a chimp have the impression that if the ape had a free choice in the matter it would much rather sneak off to some quiet spot where it could eat its prize in peace. But, as with the baboons, the ill-fated prey usually brings other chimps scurrying to the scene by its screams. In any case, many kills are cooperative affairs, and so there is no opportunity for privacy for the successful hunter.

Once a dead animal is in the hands of a chimp the lucky possessor is besieged by males, females, and juveniles, all eager for a taste. But, unlike the baboons, there is very little ill-temper displayed: the onlookers beg for a share. Begging is a varied business, and may involve gazing into the eyes of the individual with the meat, whimpering, touching the carcass as it is being gnawed by the possessor, or holding out the hand palm upwards in a gesture uncannily like a human begging gesture: the similarity of this gesture between humans and our closest biological relatives is too great to be merely a coincidence.

Although begging for meat by chimps is not always rewarded with a morsel of the kill, the success rate is clearly high enough for the behaviour to persist in its relatively peaceful form. One fascinating feature of chimp begging and sharing is that the more dominant animals rarely pull rank: in the Gombe troop one observer saw the Mike, the top-ranking male, beg for more than two hours from a lowly individual in his troop, but had eventually to retire empty-handed. Chimps never share their normal everyday plant foods, so why they share meat is something of a mystery. These amazingly human-like apes live in relatively relaxed societies, much more so than baboons', and sharing of a prized possession may be one way of maintaining a quiet life: chimps are in fact very skilled at easing social tension.

Why this difference between chimps and baboons? The answer almost certainly rests with the diametrically opposite behaviour of the adolescent males: when sub-adult male baboons near maturity they leave their troop and seek residence in a neighbouring troop (this is a very common pattern in primates); by contrast sub-adult male chimps stay where they were born and it is the females who migrate elsewhere. The consequence is that in a troop of baboon the males are not related to each other whereas chimp males are relatives of each other; it is the female chimps who are strangers. It is

therefore not surprising that baboon males are in open competition with each other: their size relative to the females, their
long sharp canines, and their frequent social aggression all
testify to this competition.

Although chimp males are bigger than their females, the
difference is not dramatic (the difference is about the same as
that between male and female humans); although chimps'
canine teeth are relatively prominent, they are not the extravagant social signals displayed by baboons; and, as we said, peace
rather than tension is the hallmark of chimp communities.
(Male chimps in fact indulge in a rather subtle form of competition with each other: they have enormous testes. Although
their bodies weigh only about a quarter of a male gorilla's,
their testes are three times as big as their ape cousins'! It seems
that, although chimps don't compete intensively with their
fellows in their efforts to mate with a female, each male tries to
improve his chances of making sure that the multiply-
inseminated female bears his offspring by depositing huge
amounts of sperm in her.)

So, if sharing of any commodity is to develop in a complex
social group, it is much more likely to emerge in the relatively
egalitarian society of chimps than among the much more
hierarchically structured open-country baboons. Were our
ancestors six to eight million years ago more like chimps than
baboons, living in groups in which the males were relatives
while the females were not? We shall, of course, never know,
but the chimp-like society certainly seems to be one from which
a cooperating hominid operating a food-sharing economy is
more likely to emerge. And it is perhaps not superfluous to
comment that in non-agricultural people, it is usually girls
who leave their gathering and hunting band to live in the band
of their mate.

Most primates are gregarious creatures: they live in groups
and enjoy a rich social life. We are, of course, assuming that the
early hominids were no exception. The benefits of socialising
that accrue to other primates as they trade off the advantages
of safety in numbers (defence against predators) against making
sure there is enough food to go round (optimum resource
exploitation) must also have accrued to our ancestors.

You only have to spend a morning quietly watching a troop
of chimps or baboons to realise what a complex business group
living is. Compared with a tranquil herd of grass-chewing cows,

a troop of primates is vibrant with subtle – and often not so subtle – social interactions: consort pairs attentively groom; adult males vie with each other for the attentions of a female just coming into oestrus; other males challenge each other for social status; mothers keep watch over their frolicking offspring; alliances form between brothers or between 'friends' in order to gain an unbeatable advantage over a temporary adversary. The scene is complex indeed, and much more so than meets the eye. Imagine, then, how much more intense social relationships must have become among our ancestors as they began to evolve their food-sharing economy: the division of labour in the dual-approach economy, combined with the reciprocal distribution of plant and animal foods, must have imposed enormous strains on our ancestors' abilities to cope with the exigencies of being part of a closely cooperating team. Undoubtedly, these were major influences in shaping the evolution of the human mind.

Viewed from a strict standpoint of evolutionary biology, living in groups is an uneasy compromise between the essential selfishness of the individual and the advantages of sharing resources with others. For many years biologists used to think of the behaviour of animals in terms of 'the good of the species' or 'the benefit to the group'. For instance, people looked upon the evolution of warning calls in birds as a mechanism designed for the preservation of the species. In fact, evolution is a very selfish affair: it progresses through benefits to individuals, not to groups. This is a very complicated argument, but it is one that is pertinent to the emergence of that unusual animal, *Homo sapiens*. Given the fact that evolutionary forces are supposed to make animals behave in a selfish way, we have to understand why sharing became such an important part of human social life.

Suppose, for a moment, we consider a group of hypothetical animals living together and steadily producing offspring. Suppose, now, that one of the offspring is born with a mutation in its genes that, for some reason, makes it twice as fecund as its fellows. And suppose that this new gene is carried through into subsequent generations of individuals. As time goes on, each generation will have a higher and higher proportion of descendants from that original mutant individual. In terms of evolutionary biology, it has been more successful than its fellows. And in the evolutionary arena, the animals best equipped to exploit the resources available to them and to reproduce most successfully are those that populate the world in the future.

The bases of all life, of course, are those long thread-like nucleic acid molecules that make up genetic material: it is the genes that determine what animals look like and how they behave. And, as parents pass their genes to their offspring who in their turn pass them on to following generations, genes in a real sense are immortal.

Genes are what make us what we are, and they are the engines of evolution. Nor is it totally unfair to regard bodies, whether of humans, elephants, or ants, as mere vehicles for propagating the immortal genes. In a recent book called *The Selfish Gene* Richard Dawkins, an English geneticist, described the role and 'motivation' of genes in the following dramatic way: 'genes swarm in huge colonies, safe inside gigantic lumbering robots, sealed off from the outside world, manipulating it by remote control. They are in you and me; they created us body and mind; and their preservation is the ultimate rationale for our existence . . . we are their survival machines'. In any case, this kind of perspective does help to focus on the qualities of particular genes when contemplating the evolution of certain types of behaviour in social animals.

Just over 20 years ago the great British biologist J. B. S. Haldane was sitting with some of his colleagues in his favourite public house near to University College, London, pondering on the problem of altruism. He took out an envelope, scribbled a few rapid calculations, and announced that he would be prepared to lay down his life for two brothers or eight cousins. As it happens, he wasn't contemplating any such dramatic act; he was working out what, in the framework of genetics, would be evolutionarily sensible. The point is that one shares half one's genes with a brother or a sister, while there is only one-eighth genetic overlap between cousins. So, as far as family genes are concerned, it is worth losing a J. B. S. Haldane if in the act of dying he saves the lives of two of his siblings or eight of his cousins.

The biological community has only recently looked seriously at these ideas, and they are currently causing something of a revolution. It is now possible to scrutinise a great deal of animal behaviour and see in it actions by individuals that are designed to benefit their close relatives. Sacrifice is, of course, not the only type of behaviour that comes into this category: anything that benefits your brother or sister (or rather more specifically, their ability to reproduce) indirectly benefits you (or rather your genes, as they are half shared with your sib). There is one sensible proviso, of course: in helping your brother or sister, you should not inflict too much cost on yourself.

It is therefore not surprising, given the commonality of genes, that families of brothers, sisters, parents, and so on, should be keen to help each other. Even the animal who utters an alarm call when it sights a predator, thus potentially putting itself in danger by attracting the predator's attention, is helping itself (or rather its genes) because among its companions will be its brothers and sisters. For obvious reasons, biologists have given the name kin selection to the evolution of behaviours that are designed to help relatives. But the story doesn't stop there. It gets more complicated, and the complication goes under the name of reciprocal altruism, something that we can immediately recognise in much of human life.

At its crudest, reciprocal altruism can be put like this: if you see someone drowning in a river, and if you can swim, then you jump in to save him. You won't be contemplating it at the instant of rescue, of course, but you'd expect that if at some time in the future your life was threatened in some way, the person you'd saved would return your favour. Reciprocal altruism works on the tacit assumption that favours are repaid in roughly the same measure. This view of apparently selfless behaviour may seem to take the altruism out of altruism, and in a sense it does.

Throughout evolutionary history animals have advanced through creating advantages for themselves over their rivals, their rivals being all members of their species. We are not suggesting intensive scheming by individuals in order to outwit their fellows: evolution works passively, and if, by some genetic quirk, an individual just happens to be better endowed than others around it, then it will thrive better too, and leave more offspring; the forces of natural selection favour such quirks. The term rival can be viewed in the context of who leaves most genes in the next generation.

Reciprocal altruism thrives in a population of 'rivals' only if everyone 'plays fair'. A chance genetic mutation may produce an individual who accepts the generosity of his fellows, but does not repay them. This kind of 'cheating' (again passive, not deliberate) may be successful for a while, but natural selection is sure to develop in the altruists the ability to detect non-reciprocators. Cheaters will soon be found out and in the end no one would be prepared to help them.

The ideal breeding ground for the evolution of reciprocal altruism is in a group of long-lived, egalitarian, social animals who remain close together throughout their lives. This means that altruistic acts can be repaid over a long period of time. You would not expect this type of behaviour to emerge in creatures that rarely encountered each other, through whatever circumstance: there would simply be no opportunity to have a debt repaid.

The description of creatures in which reciprocal altruism is likely to evolve through natural selection fits very closely the picture we have of our early ancestors. It fits the social pattern of many of the higher primates too, but as we said earlier, when the embryonic *Homo* ancestors invented a food-sharing economy, they raised the potential for reciprocal altruism to an unparalleled level. Would a female have readily shared out her cache of tubers and fruits to the hungry males if she were not sure that in return she would receive some meat when it was found? No. A *pure* altruist who always gave and never received would have fallen by the wayside in the march along the road to humankind.

As social and economic life steadily advanced through the millions upon millions of years, so the network of beneficence and obligation would have become more and more tightly knotted. It was not just food that changed hands with the covert understanding that at some future date the current receiver would become the giver, and the giver the receiver. In the mutually dependent social groups, help would be given, and expected, in every sphere of activity, from babysitting for another's child, to sharing the burden of fashioning someone else's arrowheads. People help each other all the time, and they are motivated to, not by repeated calculations of the ultimate benefit to themselves through returned favours, but because they are psychologically motivated to do so. This is precisely what one would expect: over countless generations natural selection favoured the emergence of emotions that made reciprocal altruism work, emotions such as sympathy, gratitude, guilt, and moral indignation. Indeed, passions among modern gatherer-hunters are raised most rapidly and stormily when someone is discovered to have committed some kind of injustice, however small. The whole band may become involved in the outburst, and the miscreant often finishes up in tears: the shame of being seen not to be fair is sufficient punishment.

In most higher primates, such as baboons and other large monkeys, and the great apes, there is usually a keen edge of competition running through their sex lives: usually it is the males who compete for the favours of females. Sexual competition was probably an important aspect of early *Homo* too (and in modern humans for that matter, a topic we will explore in more detail in a later chapter). This element of competition between the males must have complicated to some extent the undoubted benefits of close cooperation, both as fellows in a dual approach economy, and as occasional partners in hunting trips.

Our impulse to help someone in need is therefore involuntary inasmuch as we are responding to the command of the emotion of sympathy. The keenness of that sympathy may become dulled if the recipient continues to accept the altruistic acts but persistently fails to reciprocate when the opportunity arises. As highly cultural animals we can, of course, be beneficent knowing that our generosity will never be repaid: this is pure altruism, something you would not expect to find in the lower animal world. Nevertheless, the emotions of compassion and sympathy that stir in the breasts of such pure altruists are the evolutionary products of a long history of an intensely social creature: the decision to help may be sophisticated and cerebral, but the underlying emotion is much more basic.

We are not attempting to dehumanise humans by injecting animal selfishness into apparently noble selfless acts. Humans *are* capable of pure altruism. In fact, because evolution has created for us a context of culture, we are capable of doing more or less anything we choose. Nevertheless, it would be ostrich-like to ignore the evolutionary implications of the highly developed and intertwined senses of obligation and generosity embedded in the human brain: it tells us so much about our past.

Ever since the first tentative explorations into the new gathering and hunting economy, perhaps four million years ago, perhaps longer, the unconscious and subtle forces of natural selection have been steadily favouring the development of the uniquely human degree of doing good to others. Because by helping other people an individual helps himself, keenly developed reciprocal altruism becomes a powerful force in the success of the species. A social group in which skills are pooled achieves so much more than a collection of individualists. If an individual is particularly adept at spotting the almost invisible tell-tale signs of buried nutritious tubers, let her teach others the trick. And if a man is especially skilled in manufacturing fine stone tools, let him show others how it is done, or even make tools for others to use. Provided each individual abides by the rules of reciprocal altruism, then the species is set for evolutionary ascendancy. We are human because our ancestors learned to share their food and their skills in an honoured network of obligation. And, undoubtedly, one of the special human skills that was necessary for the greater sophistication of this network was an unusually efficient method of com-munication: language. An ability to speak allows an individual to recall events in the past, however small, and to project future actions. Others may therefore be reminded very specifically of their obligations and may note promises that must be fulfilled later. The web of language therefore makes the network of social obligation ever more tightly woven.

We will never know exactly when *pure* altruism became possible. Perhaps it arose at about the same time as the human brain was capable of generating a truly human self-awareness, a quality that also carried with it the knowledge of what death means. Because such feelings are ephemeral, even in the materialistic world of the twentieth century, they vanish without trace in the archaeological record. Only with the ritual of burying with flowers or other objects can we hope to snatch the briefest glimpse into the minds of our ancestors. But altruism, pure or otherwise, leaves no footprints in the sands of time, except perhaps in the sense that communal life in camps would have been impossible without it.

One of the very few things about which we can be certain in the long human history is that, in the evolutionary race between *Homo* and the australopithecines, it was *Homo* who won. We can guess that the other hominids eventually became extinct because their lifestyles were too similar to that of *Homo* for them all to survive: the most efficient creature outlived the rest. But we don't *know*. We don't know either whether *Australopithecus africanus* and its bulkier relation *boisei* operated a primitive gathering and hunting economy. It is very difficult to see how we could ever be certain one way or the other.

But, although anatomically the hominids were very similar to each other, inside their heads were brains that, almost certainly, created a gaping behavioural gulf between *Homo* and the australopithecines.

It is with no firm evidence that we say that the australopithecines never learned the benefits of a gathering and hunting economy, and that their hands never fashioned stone tools in any systematic way. But the rules of biology suggest that this was probably the case. We know that three million years ago hominids were making tools in a systematic way and living occasionally in camps. Probably they were doing this long before then too. Time will tell. In any case, we suggest that it was primitive *Homo* who lived in those camps and who made those tools, and, furthermore, that it was only *Homo* who did so. It is virtually inconceivable that other hominids could have developed the social structure and way of life of the dual economy: the competition would surely have been too stiff.

Was *Ramapithecus* the first primitive gatherer-hunter? It's possible. In which case the australopithecines should be seen as evolutionary progeny that sought out ecological niches elsewhere in which to develop their separate ways of making a living. However, if *Ramapithecus* had been a *proficient* gatherer-hunter, it would not have evolved into primitive *Homo*: there would have been no evolutionary pressure to do so. No, until we have reason to think otherwise, it is probably closer to the truth to think of the emerging gathering and hunting economy as the evolutionary wedge that drove itself between primitive *Homo* and the other primitive hominids as the basic hominid stock speciated to fill a variety of ecological niches.

Just as Ernest Worthing (from 'The importance of being Ernest') began life in a handbag, so the gathering and hunting economy that gave issue to humankind was conceived in a carrier bag. Sharing, and a highly developed sense of altruism followed naturally on. And so too did leisure and the first affluent society.

CHAPTER 8

Intelligence, tools, and social intercourse

Ralph Holloway works in a large basement laboratory in Columbia University's department of anthropology on the west side of Central Park in Manhattan. The windows are blacked out to exclude all natural light. In one corner is a huge, deep stone sink where Holloway carefully brews any one of a dozen or more exotic teas he stores in cans and bottles – he is something of a connoisseur of the leaf. In another corner a teaching skeleton hangs morosely close to some apparatus that looks uncannily like a piece of navigation equipment. And stored in the tall wooden cupboards standing around the walls are rows of rough rubbery hemispheres: they are singularly unprepossessing, just like large lumps of plasticene a child has abandoned in play. In fact, they are a vital link with our past, for they are casts of ancient hominid brains.

For some years now Holloway has been developing techniques for reconstructing the size and shape of early hominid brains. He goes to South and East Africa where the skulls are excavated and he makes casts which he carries carefully back to his New York laboratory. There he begins the frustrating and difficult task of analysing the cast's structure using the 'navigation' equipment to plot points on the 'brain'. And, most difficult of all, he tries to interpret what that structure means: what did the brains think when they were pulsating with blood? What lifestyle did they control for their owners?

Detailed answers to these questions will be slow in coming; some will evade us for ever, as thoughts that are not converted into concrete tangible expressions are the most ephemeral products of human existence. Already, however, Holloway has made an important – it is not too strong to say dramatic – discovery: the basic architecture of the hominid brain evolved a very long time ago – for at least three million years the brains of our *Homo* ancestors and their australo-pithecine cousins have been distinctly different from ape-like brains. Once again we are forced to peer far back into the past to see the first glimmerings of humanity, not just in the way our ancestors appeared to each other, but in what went on in their heads, and also, presumably, in the way they behaved.

To hold in one's hands the fossilised cranium of a long-dead ancestor – 1470 for instance – is a strangely moving experience: this petrified link with the past cannot fail to stir emotions – simple curiosity, perhaps a strange unease – in even the least romantic breast. Holloway's pioneering techniques take us a step further: even though what we may cradle in our hands from his work is merely a rubber mould filled with plaster, the knowledge that it traces the form of an ancient brain is very disturbing indeed. This is perhaps the closest one can ever come to reaching into the depths of human history to touch the core of what makes us such special creatures – a highly intelligent brain.

It may seem an odd question to pose, but if we are to understand our evolutionary history, we have to ask *why* humans are so intelligent? Why are we so inventive and accomplished in the world of practical affairs?

The answer may at first appear to be glaringly obvious: because it makes us such a technologically successful animal. Undoubtedly, our command of technology places us at the pinnacle of the animal kingdom. But this is a circular argument. It is just as easy to argue that we are technologically accomplished *because* we are intelligent. The point we have to settle in evolutionary terms is this: was the undoubted advantage of sophisticated technology the *primary* driving force that created the intelligent human brain? Or are we technological masters of our planet as a fortuitous consequence of needing to be intelligent for other less tangible reasons? In other words, we want to know whether the intellectual skills that enable modern man to launch a spacecraft to Mars, or to write a symphony for that matter, were necessary even in a somewhat diluted form in the practical world of emergent gatherer-hunters.

Just as the notion of man's place at the centre of the universe went unchallenged for centuries, until first Copernicus and then Darwin came along and dislodged this anthropocentric view of creation, so too has our extreme intelligence been accepted as so self-evidently paramount in human evolution that few people have bothered seriously to ask the question, *why?* The answer, as it turns out, may not be so glaringly obvious after all: it may be that during our evolutionary career we had to sharpen our wits, not so much to overcome the technological challenges faced in the practical world, but rather to handle the complexities of a uniquely intricate social life.

Withdrawing for a while from our bold assertion, we will go back to Ralph Holloway's laboratory and see in some detail what his so-called brain endocasts can tell us about our ancestors.

Basically, there are three important things about a brain: its size; its overall shape; and the intricate circuitry of the nerve fibres within it. Of these three features it is the last that is most intimately responsible for guiding the behaviour of the animal: many mentally subnormal people have brains of the right size and shape; a defective circuitry is the cause of their problem. At a more theoretical level, it is possible for brains of the same size and form to govern normal behaviour of markedly divergent sophistication: if in one the wiring diagram of the nerve fibres is highly complex, whereas in the other it is much more basic and simple, then wide differences in behavioural accomplishments are inevitable.

When Holloway makes his casts of ancient brains what he is really doing is reproducing the impression that the brain left on the inner surface of the skull. He puts a little liquid latex inside the cranium and swirls it around so as to produce a thin layer all over the inner surface. When he's done this about six times the layer is thick enough (about an eighth of an inch) for him to handle. (The latex layer is sensitive to daylight, hence the reason for the blacked-out windows in the lab.) Most hominid skulls are, of course, reconstructed from numerous fossil fragments, an accident that makes Holloway's task easier. In this case he pours plaster of Paris into the latex mould while it still lines the cranium, heat treats the cast and cranium, and then dissolves enough of the glue holding the cranial fragments together so as to release the endocast. (If the cranium is intact, the latex mould has to be collapsed so that it can be removed through the hole in the bottom where the spinal cord enters the cranium, and then filled with plaster later.)

It may seem remarkable, but brains *do* leave their signatures on the hard bony skull, even though they are not in direct contact with it. Between the soft grey cortex of the brain and the hard skull are layers of protective membranes and liquid. In some animals the protective layers are relatively thin and the inner surface of the bone carries a faithful imprint of the brain. Unfortunately, not so in hominids: nature appears to have taken special care to shelter her prized creation from damage: the robustness of the intervening membranes means that only a shadowy sketch of ancient hominid brains' geography is visible on the inside of the empty craniums. Dim though the record is, a skilled eye can read important messages in it.

If you were to place a real human brain on a table in front of you you'd see that its shiny grey surface is deeply wrinkled: this outer grey layer is the cerebral cortex and it is the area in which the so-called higher functions of the brain take place; the extravagant wrinkling is a way of giving more area to the cortex. Now, you would notice that the brain is divided neatly into two halves vertically from front to back: these are the right and left cerebral hemispheres. And each hemisphere is further divided into four so-called lobes: the one at the front (the frontal lobe) is responsible for controlling movement and for some aspects of emotions; the occipital lobe (at the back) deals with sight; the lobe at the side (the temporal lobe) is an important memory store; and the parietal lobe (at the top) has a vital role in comparing and integrating information that flows into the brain through the sensory channels of vision, hearing, smell, and touch.

Basically, a brain in which the parietal and temporal lobes predominate is human-like, whereas in ape-like brains these areas are much smaller. The overall pattern between human-like and ape-like brains is, of course, similar; the difference lies in emphasis. For instance, human's memory storage and ability to integrate a complex kaleidoscope of sensory information is much more developed than an ape's, hence the difference in the respective lobes.

In his researches over the past few years Holloway has found that *all* the hominid brains he's looked at from South and East Africa have the basic human pattern; none of them is ape-like. And this includes the 1470 skull from Lake Turkana, a skull that is around two million years old. Although the ancient hominid endocasts share a distinctly non-ape-like form, there are significant differences between the brains of *Homo* and *Australopithecus* too. For instance, the frontal lobes in two-million-year-old *Homo* are broader than in his australopithecine cousins; and the temporal lobes are better developed too. Clearly, by at least two million years ago *Homo*'s lifestyle was sufficiently distinct from that of the australopithecines for it to be reflected in a greater refinement of the basic hominid brain in our direct ancestors. But the fact that a characteristic hominid brain shape had evolved such a very long time ago is remarkable in itself. What does it mean?

First, it means that whatever it was that shaped the hominid brain into its special form was probably operating way back in the Miocene era: our common ancestor *Ramapithecus* some 15 to six million years ago was leading a life that demanded a different way of thinking from that of its contemporary ape-like cousins. The business of walking around on two legs instead of four, of exploring new energy resources, and of coping with an ever more demanding social life made its impression inside the head of *Ramapithecus* as well as on its physical appearance.

When someone finally discovers a cranium of a six-million-year-old *Ramapithecus*, who will be surprised to see the shape of a distinctly human-like brain lightly sketched on the inner surface of the skull? No one. This is something of a revolution in the study of human origins because until Holloway produced his casts of ancient brains most people guessed that, in the dynamic matrix that constitutes human evolution, the brain was among the last organs to assume a human shape. Exactly when that basic shape first started to emerge, however, remains to be seen, as too does the birth of the more refined *Homo* brain. At some time in the distant Miocene there must have been brains that were intermediate between the human-like and ape-like pattern. Was the hominid brain born complete (in its basic pattern anyhow) with the emergence of *Ramapithecus* some 15 million years ago? Or did it develop steadily as *Ramapithecus* became more and more accomplished in its hominid niche? These are questions to ponder on as we have as yet no way of answering them with any confidence.

When Holloway examines the tell-tale marks on the inner surface of the skull he is looking not only for signs of the boundaries between the four lobes, but also for the overall shape of the two hemispheres. The success of his techniques therefore depends on his having either an intact cranium – an extreme rarity in the highly fragmented fossil record – or a reconstruction that accurately reflects the shape of the head when it was living. Occasionally, Holloway complains of the extremely common and irritating habit among the early hominids of dying in such a way that the cranium becomes distorted during its slow journey into the fossil record! This is a major problem for the technique and it seriously limits its scope.

Even with a perfect skull, however, the scope of the research is still restricted because, as we implied earlier, it tells us nothing of the crucial patterns of circuitry that coursed their complex paths *inside* the brain. We see that two million years ago the external organisation of *Homo* and *Australopithecus* brains were practically identical. Are we to infer that they were identical inside too? If we were to be confronted by two factories housed in identical buildings, one of which was thriving economically while the other was gradually going out of business, would we imagine that the activity in the two apparently identical factories was the same? No. And the same answer must hold for a comparison of the hominid brains.

Although the gross architecture of the early hominid brains followed the same lines, there were differences in size. For instance, *Australopithecus africanus* around two and a half million years ago had a brain just under half a litre in volume (about 450 ccs). Its bulkier cousin, *Australopithecus boisei*, had a 550 cc brain, but the difference in brain volume here simply reflects the greater work load demanded of a brain in a bigger body. Our *Homo* ancestor at this time was brainier still, with a volume close to 800 ccs (this is for 1470). As our 1470 was probably about the same size as *Australopithecus boisei*, or even smaller, it follows that he was brainier not just in brain volume but also in wit.

Throughout their evolution, and right up to their final extinction about a million years ago, the australopithecines showed no signs of significant progress in the size of their brains. And very probably their common ancestor, which they, of course, shared with *Homo*, had a brain not much smaller than that of *Australopithecus africanus*. By contrast, *Homo*'s brain got bigger and bigger: early *Homo erectus* about one and a half million years ago had brains approaching 1000 ccs; by half a million years ago this had risen to perhaps 1200 ccs; and when fully modern humans (*Homo sapiens*) evolved about 50,000 years ago, their heads housed brains with an average of 1400 ccs. During this time our ancestors grew in bodily stature too.

Simply knowing the size of a brain doesn't actually reveal much about the intelligence of the animal that owned it: the crucial factor is the size of the brain in relation to its body. Modern humans have big brains compared with the body size. But we don't top this league, as the tree shrew, house mouse, porpoise, and squirrel monkey all do better! Among the apes, however, we are the most generously endowed.

Although it is impossible to be sure about the size and weight of the early hominids, intelligent estimates allow us to make tentative comparisons between the relative braininess of early *Homo*, the australopithecines, and modern humans. Without going into data, which anyway are far from ideal because of their inherent uncertainties, it is fair to say that the early hominids, including the australopithecines, are within respectable striking distance of modern humans in the ratio of their brain and body weights: they were definitely less brainy than we are, but not *dramatically* so. The australopithecines never improved their position, whereas with a steady addition to the size of the brain (and the body at a slightly lower rate) primitive *Homo* eventually reached its current status.

If the advancement in relative braininess between 1470 and his fellows two million years or so ago, and modern *Homo sapiens* today, is not very dramatic, how do we account for the apparent behavioural and intellectual chasm that must separate us now? One factor, of course, is that knowledge and technological know-how accumulate through cultural tradition, so that the material distance that divides twentieth-century affluent societies from ancient gatherer-hunters, say 50,000 years ago, is not matched by an innate intellectual distance. We are exactly the same animal now as we were 50 millennia ago; we simply know more now. But when we are comparing modern *Homo sapiens* with pre-*sapiens* hominids, the essential point is that relative brain sizes *can* mask enormous biological differences, as we implied earlier. During the past few million years human brains have enlarged, but they must have increased in *internal complexity* too. This is what really separates us from the brain of 1470. There can never be any *direct* evidence to back up this statement. But the rising sophistication of social organisation and subsistence economy reflected in the archaeological record make it difficult to refute.

Incidentally, although the *average* size of modern human brains is about 1400 ccs, the variation is very large, spreading from 1000 ccs right up to 2000 ccs. And apparently there is no special correlation between enormous brains and the display of genius. Although Ivan Turgenev and Jonathan Swift had brains impressive in size (just over 2000 ccs) and in intellect, the talents of Anatole France flowed from a diminutive 1000 ccs brain (this is not much bigger than the brain of 1470!). These examples emphasise that it is the *internal organisation* that is most important in determining the scale of wit and intellect.

We can now begin to approach the question of human intelligence by asking, what are brains for? Again, this may seem an unusual question to pose, but it does help us obtain an insight into the evolutionary history of intelligence.

Basically, we have brains in our heads – whether we are humans, monkeys, mice, or lizards – in order to create our version of 'the real world'. Animals in the different parts of the evolutionary spectrum lead lives that are more or less complicated. If your life is very simple, such as a frog's, for example, then you can thrive from day to day with a minimum of information about the world outside. If, however, you are an African wild dog then the world you create in your head is a much richer place than that in the frog's head: it has to be because you have a keen sense of sight, smell, and hearing, and you have to cooperate with your fellows in hunting fast-moving prey. This is far removed from the solitary activity of sitting by a pond flicking a long tongue at passing flies! Little wonder then that a wild dog has more extensive wiring in its head than does the frog – it has to, otherwise it simply could not *be* a dog.

More than 200 million years ago, in the late Palaeozoic and early Mesozoic eras, the world was ruled by reptiles large and small: among them were the dinosaurs. They lived by day, and their world was largely a visual world, however limited it may have been. But, although these ancient creatures, and their modern counterparts, saw the world with their intricately structured eyes, we have to question whether they *experienced* what they saw. Almost certainly they did not, and the reason we say this is because a reptile's response to a visual stimulus is extremely stereotyped: the frog flicks out its tongue at a passing fly, or at any other moving object about the size of a fly, however unpalatable it is.

More than 20 years ago Roger Sperry, an American neurobiologist, did a classic experiment on a frog: he rotated one of its eyes so that the animal saw the world upside down. Now, if humans, or cats, are given special spectacles that invert the visual world, they soon adjust to the 'wrong' images and can move about and handle objects competently. A frog *never* adjusts: if a fly passes overhead it consistently flicks its tongue downwards.

The reptilian eye is a complicated piece of machinery, so much so that much of the essential visual analysis is performed within the retina itself: very little information passes to the brain – there is simply no need. About 200 million years ago the first mammal-like creatures evolved. They were small and they had to make their living under the cover of night. Reptilian vision wasn't much good to them, and so, through the channels of natural selection, efficient hearing evolved. Unlike the eye, the ear simply did not offer sufficient space to package the necessary analysing machinery: hence it had to go in the brain. This development was the first important step in the evolutionary growth of brains.

At about the same time the primitive mammals' visual system adapted to the dim light of their nocturnal habitats. Night vision was born, and made use of twilight hours and occasional moonlight to give useful information about distance and movement. (Most modern mammals are still nocturnal, incidentally.) Because these ancient animals had the benefit of information from their ears and their eyes, it made good biological sense for the brain to 'compare' the inputs from both sources. This so-called integration of information in the brain was not only another important step in brain enlargement, but it also meant that for the first time the world outside was coming together inside an animal's head.

Further progress had to wait until about 70 million years ago when the ascendancy of the reptiles ended abruptly and the Age of Mammals began; creatures that evolved from remnant populations of curious mammal-like reptiles began to dominate the earth. As some of the newly evolving mammals emerged from their night-time habitats and began living daytime lives they re-evolved daytime vision. Unlike the reptiles' vision, however, this modern version depended heavily on analysis in the brain. Colour vision developed too. And, more particularly for nocturnal animals, the sense of smell became very acute. All these developments were 'designed' to give animals a clearer picture of their world, and those animals that had two or more windows on the world (sight, hearing, and smell) evolved ways of integrating the information, a crucial advance in the emergence of consciousness. And the brain grew still more.

Compared with the ancient reptiles, the mammal-like reptiles equipped with their better hearing had brains four times bigger than their ancestors. And when true mammals arrived they carried in their heads brains that were five times bigger than their immediate forebears. The primates are the most splendidly endowed, with monkeys and apes being relatively twice as brainy as the rest of the mammals. The hominids, of course, are in a class of their own.

The progressive evolution of the higher primate brain therefore produced a machine that, through comparing incoming information from several sensory channels, and then setting the results against information kept in a capacious memory store, can create a vibrant image of the world outside. It can recognise objects as discrete entities rather than as simply part of a patchwork pattern. It can equate the furry appearance of a cat with a small animal that miaows (without integrating our sensory information we would be unable to do this). And it can project events into the future as well as store happenings of the past.

This is intelligence. But what is *human* intelligence? Although the human intellect is unparalleled in the animal kingdom, the special status of the human mind is gradually being eroded as we learn more and more about what goes on in the heads of the great apes (particularly chimps and gorillas).

One quality in the world that is appreciated by the human mind is the passage of time. A true perception of events in the outside world is impossible unless time is automatically built into the analysis of information about them. In this sense any animal that is dealing with a perceptual world of even a small degree of complexity must have a brain capable of comparing events in the context of time as well as in space. But the *appreciation* of time is quite separate from this. Philosophically, one could view oneself as being constant observing a changing world, rather than the other way round.

Tom Stoppard, the English playwright, refers repeatedly to such philosophical uncertainties in his brilliant play 'Jumpers'. In the closing scene the chief character, a professor of moral philosophy says that '. . . all the observable phenomena associated with the train leaving Paddington (station) could equally well be accounted for by Paddington leaving the train . . .'. We *know*, however, that trains leave stations rather than the other way round, and we *know* that it is the world that is constant and not us (at least we do when we are more than about five or six years old). But do apes?

Until recently the answer would have been, probably not. But a 'conversation' between a Stanford University graduate student and her 'talking' gorilla, Koko, should make us not quite so sure. Three days after Koko had bitten her in a fit of anger, Penny Patterson, the student, asked the ape (in sign language) 'What did you do to Penny?' Koko replied 'Bite'. 'You admit it?' Patterson continued. Looking a little contrite, Koko said 'Sorry bite scratch'. Patterson then asked Koko why she had bitten her. 'Because mad,' came the answer. 'Why mad?' 'Don't know.' The conversation ended.

This interchange is remarkable in that it shows Koko referring to events, and emotions, some distance in the past. Normally Koko refuses to converse about her mischievous acts immediately after the event. Here she was talking about something that had occurred three days previously. Exactly what kind of time awareness Koko has is difficult to say, but it is only through anecdotal evidence such as this that we can get some glimmerings of apes' time perception.

We share with the great apes much the same kinds of sensory tools for building up a picture of the world in our heads. But one tool that we alone have is spoken language. This is a topic we will explore more in the next chapter, but it is relevant here because it is quite legitimate to view language as a very special tool for sharpening up still further our perceptual image of the world outside our heads. By naming objects and categorising them, we can manipulate our ideas and impose order on to what would otherwise be an unco-ordinated kaleidscope of perception. And by speaking to each other about the images in our heads we create a genuine shared consciousness in a way that is impossible without words. The evolution of a spoken language must certainly have been influential in the expansion of the human brain during the past couple of million years. Important though language is, both as a channel of communication and as equipment for thinking, the really special feature of the human brain is its use of language to question our place in nature. Intense intellectual curiosity is a hallmark of humankind.

This curiosity would be impossible without a true conscious awareness perfusing our minds. Awareness is a philosophical quality that can provoke endless debate: we would accept that humans are aware; but is a chimp; is your dog aware, or your cat; is a frog aware; or a fish? We can say that different classes of animals inhabit different perceptual worlds because of their different channels of sensory perception and machinery for processing it. But are they aware of their worlds? Are they aware of themselves as individuals within them?

Why should any animal, including humans, be aware at all? It is surely conceivable that neurological machinery as complex as that in our heads, or in the head of a dog or a mouse, could coordinate our responses to the environment and respond to our physiological needs without raising anything to the level of conscious experience. We *could* be machines. But we are not. Why?

Starting from the acceptable premise that, because it is there, conscious awareness is biologically useful, Nicholas Humphrey, a British psychologist, suggests that it helps social animals interact with their fellows more effectively. If one is aware of one's emotions as well as simply responding blindly to them, then one can better understand the behaviour of others. If one knows what pain feels like, then one knows why others grimace when they damage themselves: without this, sympathy would be almost unthinkable. And if one is aware of subtle calculations in one's mind that are designed to win social advantage, one can deal with other people's social manoeuvres more effectively, In fact, it is arguable that, without conscious awareness, social manoeuvring' would be greatly limited.

By contrast animals that lead relatively solitary lives, untouched by the unpredictable complexities of intense social interactions, would have little need of the quality of awareness of which we speak. Exactly where in the animal kingdom awareness begins and ends is impossible to say. Indeed, the signpost 'awareness starts here' depends crucially on one's definition. Some physicists and philosophers are prepared to argue that even a rock is aware inasmuch as its molecular framework may respond to changes in its environment. But we are talking of a higher, more human, quality than this.

One important milestone on the path to full human conscious experience is *self-awareness*, the knowledge that one is an individual among many. And following close on self-awareness is *death-awareness*. Unless the strands of perception weave together to create the specific notion of one's being a separate entity – an individual – in company with other separate entities, the idea that life at some time will end is simply not tenable. When the certainty of mortality first flickered in the minds of our ancestors they must have begun to care for their dead; they invented ritual burial.

Inevitably the archaeological record is too sketchy for us to be able to pinpoint the beginnings of ritual burial. Ironically, one of the earliest and least equivocal instances of this moving human behaviour is at a Neanderthal site in the Zagros Mountains of Iraq. There, in the Shanidar cave, a man was laid to rest on a bed of flowers more than 60,000 years ago. As far as we can determine, the Neanderthal race was an ill-fated side-branch of human evolution that arose around 100,000 years ago and disappeared perhaps 50,000 years later. Although Neanderthal man failed in the evolutionary march to humankind, apparently they were human enough to take care of their souls.

More indirect evidence of a human spirit flickering in our past is a lump of sharpened ochre that was discovered in the corner of the remains of a three-hundred-thousand-year-old shelter overlooking the Mediterranean in Nice. The camp, known as Terra Amata, had been the springtime home of hominids in transition between *Homo erectus* and early *Homo sapiens*. There are no signs left of how they used the ochre, but it must have been for decorating something: their shelter, perhaps, or even themselves. Perhaps the fact that the ochre was sharpened to a point betrays the intention to use it in some kind of ritual.

We can go back a further 200,000 years, to the Choukoutien caves in China to find what at the moment is the earliest indication of ritual. There, half a million years ago, a group of our ancestors ate the brains of some of their fellows. We know that the occasion was not an ordinary meal because the participants took the trouble to widen the hole that leads into the bottom of the skull. Compared with simply smashing the skull and scooping out the brain, this is a very tedious way of getting a meal. It is redolent of ritual. Whether the event was one that followed an aggressive encounter between hostile groups, or was a mark of respect for the dead, is irrelevant to this particular argument. The important point is what the ritualistic act gives us as a brief glimpse into the primitive human mind.

Apart from elephants, which occasionally cover the corpse of one of their fellows with branches, no animal buries its dead. (We don't, of course, know what the intention of elephants is in their apparent 'burial'.) Ritual is not entirely absent from the animal world, however, for on one occasion on which Jane Goodall was observing a group of chimps at Gombe on the shores of Lake Tanganyika she witnessed a remarkable sequence of events that could claim to be ritualistic expression in embryo. The chimps were feeding in a huge fig tree with a tremendous storm threatening in the black clouds hanging overhead. As the first drops of rain fell the chimps slowly came out of the tree and walked up a grassy slope; there were males, females, and a few youngsters.

As they reached the top of the slope the clouds opened, pouring down torrential rain which was accompanied by an enormous crash of thunder right overhead. One of the big males began swaying from foot to foot, making the pant-hoot noise so characteristic of chimps. Suddenly he charged down the slope and came to rest at the foot of a small tree. He was followed by two more males, one of which grabbed a branch which he waved above his head before flinging it down the slope. The other, when he reached the bottom, rhythmically swayed a tree before he too grabbed a branch which he dragged off. More males followed, some of them wielding branches, some just swaying.

When all the males had been through their display, the individual who started it all plodded back to the top. He was followed by his fellows, and they flung themselves into the whole display once again with equal enthusiasm. In fact, the cycle continued for more than 20 minutes, to the accompaniment of claps of thunder and a sky lashed by lightning. The females and youngsters sat in the branches of a tree and watched. It must have been an awesome sight. Jane Goodall called it the rain dance.

What was going on in the minds of those animals – our closest relatives – remains their secret, but it is hard to think of their display as anything other than a primitive ritual in response to the drama of the elements.

In the less dramatic and more controlled atmosphere of the psychological laboratory, the ability of various higher primates to recognise themselves as individuals has been put to scientific test. Bearing in mind the tantalisingly subjective quality of awareness, we should not now be too surprised to discover that chimps pass this test: they can recognise themselves in a mirror. Without a *self-identity* there would be no possibility of this kind of *self-recognition*. Orang-utans also appear to have taken heed of the Delphic command: 'Know thyself'. It will be surprising too if gorillas fail this test of awareness. Beyond this, however, the rest of the primates see only another individual when faced with their reflection in a mirror: gibbons, mandrill and hamadryas baboons, spider monkeys, capuchins, several macaques, all appear to fail to recognise their own image as themselves as judged by scientific experiment.

Awareness here is a question of degree. As social creatures (for the most part anyway) these animals who failed the mirror test almost certainly have a higher level of awareness than, say, a frog. That their awareness doesn't stretch to the expression of self-awareness, of self-identity, while in chimps and orangs it does, reminds us perhaps that the human mind is at the extreme of a spectrum of biological qualities, and not, as some people would have us believe, the product of a special creation.

We will return to the theme of social pressures on the emerging human brain later in the chapter. But now we will take a brief excursion into the world of practical affairs, for it is here that an animal's style of behaviour shaves close to the cutting edge of natural selection: you fall by the wayside of the evolutionary path if you fail in the practical world, even if you are an adept socialiser.

For a very long time scientists have focused on tools as a mark of human evolutionary ascendancy. Humans, it was said, were the only animal that had learned the trick of using tools in order to earn a living. This would have been a very neat division between ourselves and a technologically bereft animal kingdom. But unfortunately it isn't true: stories about birds, sea otters, and even molluscs started to turn up, thus trespassing on this supposed human speciality.

For instance, the Galapagos woodpecker finch, as well as some other birds, prises insects out of small crevices with small twigs held in the beak. Egyptian vultures hurl stones from their beaks at ostrich eggs to break them open. There is a small sea snail that lives off the Pacific coast of California that piles small stones on one end of its foot to right itself after it is tipped over. And at least one sea otter appears to have adapted its habits to the twentieth century. These fast-disappearing animals often feast on abalone that they fish from the sea bed. They float to the surface, lie on their back, and then crack open the shellfish with a stone. The late William Bishop, a British geologist, used to tell of a sea otter he saw one day swim to the surface, lie on its back, and crack open the abalone – with a coke bottle!

When the human preserve of tool-using had been trespassed sufficiently to scupper this particular notion its proponents retreated to *tool-making* as the human speciality. No other animal actually made tools, it was argued. That is until someone noticed that they do: our close cousin the chimpanzee makes a number of implements for improving its exploitation of a number of different kinds of food, including ants, termites, honey, and the brains of slain animals. Now that chimps have forced us to abandon this piece of technological territory too, we are left with the less than dramatic claim that humans are the only animals who *use tools to make tools*.

Tools are important in evolutionary biology if they help you make a better living with greater ease. By comparing the termite-eating habits of the chimps and baboons at Gombe we can obtain a very clear example of the benefits to be had of using tools. The termites live in huge numbers in colonies that can raise large clay mounds which the wind and the rain sometimes sculpt into the shapes of fine fairy castles with delicate turrets. Every year, for a period of about two weeks at the onset of the rains in October or November, the winged termites migrate. The termites seem to take as a signal to migrate the beating of raindrops on their castle walls. (This cannot be the only signal, however, because the termites prepare for the migrations *before* the rain actually falls, and if the rain fails to come the castle's inhabitants eventually leave without the call of the beating drum.)

It is during the termite migrations that the baboons seize their opportunity. The agile primates leap up and down the mounds catching the winged termites as they fly away or as they crawl out of their newly opened tunnels. The baboons catch their prey with their hands or their lips, a practice that is obviously rewarding enough to keep the animals at their task for long periods of time, but which cannot be greatly productive in terms of the bulk of food they eat.

Chimpanzees do much better. They too wait for the rains for their termite season, but they approach the task much more methodically. When they go on a termiting expedition they break off a short flexible twig or grass stalk, sometimes before reaching the mound, sometimes when they've arrived. They then scratch away the opening to a tunnel, insert the twig so that it follows the twists and turns of the passageway, hold it in position for a few seconds, and then withdraw it. Usually the tool is festooned with soldier termites biting furiously at the unwelcome intrusion to their colony. The chimp then picks off the termites with its lips and crunches the victims before they can inflict a painful bite on their captor. Chimps sometimes fish for termites for more than two hours at a time, and sometimes they do it in the company of other chimps.

By breaking into the closed tunnels and fishing with a probe chimps both improve the yield of each termiting expedition and extend the termiting season to about a month (double that of the baboon).

Termites also provide a tasty meal for people in East Africa, and, employing subterfuge as well as technology, they exceed the productivity of both baboons and chimps. People often use larger probes, sometimes driving poles into the termite mounds. But the subterfuge they use is to beat on the mound with sticks, a trick that encourages the termites to leave their home as if it were raining. In western Zaire the same effect is produced by making a slapping noise with the tongue.

Chimps also eat driver ants (sometimes called safari ants), a meal that demands different technology and a different strategy from that of termite fishing. These much-feared ants, which live in colonies with as many as two or three million individuals in them, stream in columns until they find a suitable spot to construct their underground residences. Their teeming colonies seem able to penetrate almost anywhere and destroy almost anything they choose. Young livestock – and human babies – have to be guarded from their vicious biting armies. (When he was just a few weeks old Richard's elder brother, Jonathan, was attacked in his cot by a migrating army of these ants, but he was snatched from their jaws before they did any real damage.)

Because safari ants are well equipped to defend themselves, chimps have to take great care when they go ant dipping. The tool they use must be long, straight, sturdy, and smooth: as with termite fishing, chimps sometimes select and make tools before they set off in search of ants or they may break off a suitable branch when they come across either a colony on the march or the opening to their subterranean home. The chimp strips the leaves from the branch, and also removes the bark if it is particularly rough; this is very important. It then dips the probe, which is usually about two feet long, into the moving colony and waits for the soldiers to march about three-quarters of the way up. Then, in a swift, deft sequence of movements the ape swings the tool upwards into a vertical position; puts his free hand around the bottom of the tool; with this hand he sweeps up the tool, scooping the ants on the way; the writhing mass of ants are then popped immediately into his mouth whereupon he disposes of them by gnashing frantically with his jaws before the soldiers can organise themselves for an attack on the soft flesh of the chimp's mouth.

If the chimp finds an underground colony rather than a migrating column he first has to scoop out the entrance, either using his hands or with the end of the tool. One can see how very close this digging comes to human excavation of nutritious tubers with the aid of digging sticks. Close, but not close enough: chimps never dig for underground plants.

Chimps also use sticks for prising the nests of arboreal ants away from tree trunks. And sticks are useful too for dipping into beehives to reach the honey. Chimps are certainly the most resourceful animals in the world of technology: they make sponges by chewing a bunch of leaves with which they can soak up water from crannies in tree trunks, much as the !Kung people do; and they smash open hard fruit or nuts by pounding them with rocks, sometimes using an anvil and a hammer. All of chimps' subsistence technology is related to plant foods and insects. They occasionally use a leaf sponge for soaking up the brain juices in the skull of an animal they've killed but they never use sticks or stones either for killing or for sharing prey. This should give us pause for thought when we are considering the technology and lifestyle of the very early hominids. Digging

144

sticks and prehistoric carrier bags would be as archaeologically invisible as are chimps' digging sticks and leaf sponges.

By contrast with their evolutionarily more advanced cousins, baboons are not very technologically minded. They occasionally attack hard fruit in the way that chimps do; and they have also been seen to search for scorpions underneath stones which they then use to dispatch the creature before eating it (being careful first to remove the sting). And Jane Goodall and Craig Packer once saw an olive baboon use a stone as a napkin for wiping a sticky fluid from its face. But they never make tools in the way that chimps do.

In many scenarios of the early stages of human evolution technology has rapidly assumed the guise of weaponry, making our ancestors 'the killer ape' (Raymond Dart's words). But, as we have said, although chimps are adept at making tools and even more skilful in using them, they never direct this technology to getting meat. However, in the occasional antagonistic confrontations between the chimps and baboons at Gombe, the apes often hurl stones at the monkeys, sometimes with a respectable aim. The baboons usually flee at this point.

Chimps frequently brandish branches too, but this is more to impress each other in manoeuvring for greater social status than in malevolent intent. Mike, a male in the Gombe troop, once put a couple of Jane Goodall's empty kerosene cans to use in simian politics by kicking them along the trail at great speed, creating a tremendous din. So impressed were his fellow apes that he immediately rose to the top of the social ladder. Clearly, for Mike, wit rather than gleaming canines or rippling muscle was more effective in gaining social advantage. And this must hold for other higher primates too. For instance, one of the Pumphouse gang baboons at Gilgil is a magnificent male called David. Although he is still not fully mature, he is physically very impressive, both in size, coat, and canines. And he is obviously anxious for social success. He is able to terrorise much of the Pumphouse gang, simply by displaying his physique and threatening. But, as yet, he is *not* socially successful: he often fails to make alliances with other baboons in the troop, and he is not very good with the women. In fact, he's a typical immature macho who is overimpressed with his physical abilities and has not yet discovered that intelligence, not muscle, is the key to social and political advancement.

The greater part of the skill in subsistence technology, in

termiting, for instance, is in *using* the tool rather than in *making* it in the first place. As American anthropologist Geza Teleki discovered to his embarrassment, chimps make termiting look deceptively easy. In an attempt to discover what clues chimps use for locating tunnels in termite mounts, inserting the probe, and encouraging the soldiers to bite, Teleki tried it for himself. Not only could he find no sign of tunnel entrances without hacking at the mound with a knife, but he also failed to attract more than half a dozen termites to his probe. He said his experience '. . . left me with a healthy measure of respect for chimpanzee technical ability, as well as with a nagging suspicion that the physical and psychological capabilities needed to develop, apply and transmit such skills may differ in degree but not in kind from those needed by humans to locate, expose, and gather insects and subsurface flora.'

As Teleki implies, there is more to tool-using than simply making and using tools. There are cognitive powers required to associate the manufactured tool with suitable raw material, as well as with the implement's application to a particular job. In other words, you need more wit to see that a twig festooned with leaves could be used for ant dipping by stripping off those leaves than you do to recognise that a smooth straight piece of grass will serve in termite fishing: one has to be converted into a tool, the other serves as a tool without modification. Visualising a tear-drop shaped handaxe in a formless lump of rock goes an intellectual stage further, as the product bears little relation to the raw material. Only humans manage to make this last intellectual leap.

So, chimps have to be bright to make and use the tools they use in their subsistence technology. But, as every psychologist knows, chimps are *very* bright. During the past few years our simian cousins have mastered many tricks and have accomplished with ease intellectual leaps through psychologists' hoops, all of which betrays extraordinary cognitive abilities. They often achieve respectably human scores on standard psychological tests, they work out the correct routes on complex mazes that have undergraduate students scratching their heads, and they can successfully solve multi-step sequential problems. And, of course, they can learn American sign language and converse with computers using a richly expressive code. Chimps indeed are *very* bright. So too are gorillas and orang-utans.

The question is, how intelligent do you have to be in order to recognise the opportunity of better exploitation of termite mounds as a source of food, and then construct and use an implement to do so? The answer, surely, is that chimps are far more intelligent than they need be for these relatively rudimentary tasks. Their talents are wasted. Or are they? Are we to assume that evolution has produced in the heads of chimpanzees – and gorillas and orang-utans too – brains whose powers far outstrip the use to which they are put? That indeed would be a waste, and indeed it would be uncharacteristic of the usually economic forces of evolution. Some American and British biologists have been pondering this apparent evolutionary profligacy and they have come up with the notion that, yes, chimps and their fellow apes do have to be as intelligent as they are, but for social rather than technological purposes. Nicholas Humphrey has developed the idea most fully.

During the evolution of mammals, carnivores had to become particularly intelligent because seeking out, stalking, and killing prey is more demanding than standing about eating grass. (These grass-eaters, the ungulates, become brighter too, probably as a means of trying to outwit their predators!) Although the exigencies of the practical world must have exerted some evolutionary pressure on the brain power of the higher primates too, these animals really had to sharpen their wits in order that they could cope more effectively with each other.

A chimp troop is a dynamic entity, an ever-changing kaleidoscope of practical matters and social moods. The practical world itself is relatively predictable, certainly so compared with the frequently erratic actions of the individuals within it. Although considerable cognitive skills are undoubtedly required for exploiting a diverse and widely scattered food resource, they become relatively basic when compared with the intellectual demands of. making and maintaining social alliances, of political manoeuvring to gain subtle advances in social status, and of simply interacting with another essentially unpredictable individual. The behaviour of plants and prey is more or less certain; the behaviour of chimps and humans in complex social organisations is not. You need a keener wit for dealing with ever-changing relative uncertainties than you do for coping with relative certainties. And, as we suggested earlier, it helps to be consciously aware of the world in which you live if it is populated by individuals with whom you constantly interact.

The argument then is that the pressures of social life were an important engine in the evolution of intelligence, both in the higher primates and ourselves. Technological and other subsistence demands must have played some part too, simply because a basic technology does confer economic advantages. But that technology need be only very basic for it to be very effective: chimps' termiting techniques double the time they can feast at the termite mounds, and a very simple carrier bag is all that is required to establish a reciprocal food-sharing economy.

Between the period three million to one million years ago hominid stone-tool technology advanced from a simple tool-kit of crude cores and flakes to the more sophisticated Developed Oldowan and Acheulian which had upwards of a dozen discrete tool types. Great though the progress was, it could hardly be said to have reached the peak of high technology. And yet during this period the size of our ancestors' brains virtually doubled! Between one million and a quarter of a million years ago stone-tool technology continued to progress at its snail's pace. And again this was accompanied by brain expansion of about a third. Although the ingenuity that undoubtedly went into the application of this basic technology probably provided its own evolutionary boost to brain growth, as did the psychological demands of the mechanics of gathering and scavenging/ hunting, it was surely not enough to account fully for the remarkable emergence of human mind.

A crucial part of operating the mixed economy of gathering and hunting, apart from the mechanics of knowing where and when to find food, was the intensely enhanced social interaction. Being part of a group cooperating in different ways to achieve the same goal can be a very frustrating business, as anyone who has ever served on a committee certainly knows! Restraint, persuasion, tact, submission, aggression, perception, and a good sense of humour, all play their part in successful cooperation. And on top of all this is the desire – conscious and unconscious – to ensure that the system of reciprocal altruism operates as it should, with no one gaining unfair advantages.

Because the potential benefits to the individual of reciprocal altruism are great, we can be sure that the galaxy of emotions that underlie the system have been evolving in us for a very long time. For instance, the tendency to like people who are not necessarily closely related is essential if such altruistic bonds are to remain intact: we are usually altruistic to people whom we like, and we usually like those people who are altruistic. But when someone refuses to reciprocate, the 'injured' partner is suffused with what American biologist Robert Trivers calls moralistic aggression: anger at such outrageous behaviour. As we said in an earlier chapter, modern gatherer-hunters explode into violent passions when such injustice is perpetrated. The effect of moralistic aggression is two-fold: first it prevents any further altruism to the non-reciprocator; and second, it usually jolts the non-reciprocator back into reciprocating. The corollary of moralistic aggression, of course, is the emotion of guilt in the wrongdoer.

The emotions of sympathy and gratitude also serve to regulate reciprocal altruism. Faced with someone in need, one feels sorry for them, and the greater the plight the deeper the sympathy, so the more one is motivated to help them. When the recipient receives help to relieve his plight he feels gratitude towards his helper, and this is a psychological motivation to reciprocate in the future. The interaction of these two emotions is actually quite complex, and some sociologists suggest that the tendency to reciprocate depends on many factors. For instance, if someone is helped out of big trouble by a friend, rather than simply out of an irritating inconvenience, then he is much more likely to repay the kindness. And if helping someone costs a lot of time or effort, then again the recipient is more likely to reciprocate.

Once the system of reciprocal altruism becomes embedded in a social animal it can quickly become highly complex, especially in an animal that communicates by a spoken language. During the course of natural selection individuals are certain to arise who (unconsciously) try to 'cheat' in the altruism game. They may produce convincing sham moralistic aggression, sham guilt, sham sympathy, and sham gratitude in an 'attempt' to take more than they give, a situation that could be biologically beneficial in the short term at least. Although an animal without spoken language could indulge in such 'cheating' it is much more effective via the spoken word.

Just as natural selection inevitably produces would-be cheaters, it will just as inevitably give rise to individuals capable of detecting cheating. And so the game of bluff and double-bluff begins, with the new emotions of trust and suspicion being invented.

During its early evolutionary career, human reciprocal altruism, and the layers of emotions and counter-emotions that go with it, would have been an integral part of our ancestors' 'natural' behaviour. As we all know, at some time in our history consciousness in the human mind rose to a level sufficient to allow freely given altruism with no expectation of return, and equally freely perpetrated and deliberate cheating. But the underlying framework of emotions and motivations are deeply embedded in the human mind as part of a system that must have contributed significantly to the unusual evolution of the human brain over the past three and more million years.

Although there can have been no *single* force responsible for the extreme development of the human intellect – evolution rarely works in such a monolithic way – we can be sure that the demands of social intercourse provided a major thrust in the growth of the human brain. The intellectual exigencies of a gathering and hunting economy, and the accompanying advantages of technology, must also have played their part. What the world can now create in our heads and, because we are human, each of our worlds is different in subtle ways from everyone else's.

Over the past million years human social interaction, and probably technological innovation, too, rose to such a pitch that it generated its own stimulus for further evolution. As we more and more began to shape the world around us, to impose our will on it, we gradually became a truly cultural animal. Our ancestors invented culture, and as it grew stronger and richer it provided a unique environment which was to nurture the human mind to the point at which we know and experience it today. This is the next chapter of our story.

CHAPTER 9

Language, culture, and social psychology

For six years Keith and Cathy Hayes tried patiently to teach Viki to talk, praising her when she seemed to be making progress, comforting her when she became distressed at her failure. During those six years Viki learned to do more or less everything else a normal infant does – and she was just as mischievous. But she never learned to utter more than four words – mama, papa, up and cup – and this only with great effort. In the end the Hayes gave up, defeated and dispirited. They had to accept that Viki was never going to learn to talk properly.

The problem was not that Viki was mentally subnormal: she was bright and alert in every way, in spite of her inability to talk. She was a chimpanzee.

When they set out on their heroic but ill-fated experiment more than 20 years ago the Hayes were hoping to succeed where another researcher, William Furness III, had failed. At the beginning of the century Furness nursed and nurtured a young orang-utan in Borneo, and he, too, painstakingly rehearsed the ape in the ways of human language. As with Viki, the young orang amassed a miniscule vocabulary of words which, to the sympathetic ear, sounded like papa and cup. In reporting his limited achievement to the American Philosophical Society in 1916 Furness described how one day he had thought to introduce his orang to the joys of the swimming pool – apes generally dislike water intensely. He carried the ape into the pool, whereupon she became terrified: 'she clung, with her arms about my neck, kissed me again and again, and kept saying "papa! papa! papa!" Of course, I went no further after that pathetic appeal'.

These abortive attempts to teach apes to speak with a human tongue appeared to confirm René Descartes's assertion

of more than three hundred years ago that the human possession of a rational soul carves out an unbreachable chasm
between us and the rest of the animal kingdom. The principal
feature of our rationality is a propositional language, the
'faculty of arranging together different words, and composing
a discourse from them'. Animals are machines, he said, mere
automats; while in the human brain there burns a true spirit
kindled by the flame of a spoken propositional language.

It remains true that, apart from the mechanical and repeti-
tive utterings of parrots and myna birds, no animal can
speak a human language. And it is also true that what Plato
called 'the loom of language' must have been the primary
implement in weaving together the richly patterned fabric of
human culture. The question we have to pursue once again is,
Why? Why is it that during the evolutionary journey from a
speechless ape-like ancestor we learned the trick of stringing
together complex sounds in a meaningful way?

Many animals make complex sounds, of course, but only in
humans do those sounds represent objects or events in an
arbitrary yet symbolic way. These sounds – we usually call
them words – are *inventions* of the human mind: a tall, thin
brown structure with branches and leaves is called a tree,
not as a result of divine instruction in the way that some
ancient greek philosophers used to view the origin of words,
but as a product of English cultural tradition. Words are
arbitrary inventions inasmuch as they are meaningful names
only within a particular culture: just as there would be no
culture without language, there would be no language without
culture.

Even though an Englishman may call those thin, tall brown
leafy structures trees, a Frenchman says they are *arbres*, while
a German uses the word *baum*. But even though native tongues
and dialects employ different sounds for the same objects or
actions, the rules we use to couple these sounds together in
conversation with others or in private thoughts in our heads
are fundamentally similar in every part of the globe. Within
the rich diversity of tongues throughout the world there is a
profound commonality of human language. So deeply embedded in the human mind is the capability of language that,
unless a child is reared in total isolation, it is virtually impossible
to prevent the infant from learning to speak. Various Egyptian
Pharoahs, Moghul emperors, and Scottish kings have in the

past performed cruel experiments in which children were brought up in a world deprived of words, the experimenters' aim usually being to demonstrate not only that the ability to speak is innate but also that the fundamental language of the human race coincided with that of the country over which the noble experimenter reigned! Although the rulers' chauvinistic fantasies went unsupported, most psychologists now accept that the human brain is particularly attuned in infancy to acquiring spoken language.

We may talk to our dogs and our cats, even to our rhythmically gaping goldfish and our immobile house plants, but they will never answer back in like manner. Human language is useful only for communicating with other humans, and then only within a specific linguistic culture. So, perhaps this is the answer to our question of Why?: because it helps us *communicate* with each other more efficiently. If 1470 and his friends had been able to discuss *in detail* their plans for the following day's search for meat along the shore of Lake Turkana, for instance, they would surely have prepared more than competitors who communicated with each other by the more limited channels of ape-like grunts and gestures. But, as with the evolution of human creative intelligence, more effective communication through spoken language may simply be the fortuitous by-product of our ancestors' need for words for other less obvious reasons.

Perhaps the pressures of natural selection planted in the heads of primitive *Homo* the ability for spoken language, not so that one individual could tell another what he should be doing next in the practical matters of daily life, but so that everyone could *think* more effectively about the world in which they made a living. Perhaps too the complex operation of an advanced gathering and hunting economy would be impossible without a series of *social rules* that could be elaborated and transmitted only through the medium of the spoken word. We will therefore focus on human language from the perspective of personal and social psychology: if Plato saw 'the loom of language' as a tool, then we suggest that it is more a tool of individual imagery and group cohesion than simply an implement of communication.

To the 'Why?' question of language we have to add 'When?'. Did 1470 talk to his relatives and friends, 'talk', that is, in a web of organised utterings that was significantly more sophisticated than simian sounds? Or did the birth of spoken language have to await the arrival of *Homo erectus* around one and a half million years ago? Maybe real language is a relative infant in the mental evolution of humankind, arriving on the scene perhaps a mere 50 millennia ago as the last ingredient of modern *Homo sapiens*? If we had to rely on direct evidence sketched on the path to modern man then we would have to conclude that words were invented just a few millennia ago: the earliest examples of recognisable writing appear about 5000 years ago with the Sumerian civilisation. No one, however, would argue that human language has so short a career. And yet, apart from writing, there can be no other *direct* record of this essentially ephemeral product of the human mind. So how are we to decide whether or not 1470 conversed with his friends in a style that we could concede as being a spoken language?

That spoken human language *did* emerge from the melting pot of hominid evolution is unquestionable. But, because of its nature, *why* or *when* it arose must remain forever the secret of times past. Commenting in typical sardonic fashion at the first major international meeting on human language origins held in New York in 1975, Ralph Holloway said, 'If there is any hallmark that might be said to be unique to the human animal, it is surely the ability to speculate about the origin of language. I have too much respect and regard for the intellectual capacities of chimpanzees to imagine them prone to such exercises in egoistic futility.' Although the exercise is in a real sense scientifically futile, because there can never be an answer, the inexorable nature of human intellectual curiosity is such that we are bound to continue asking the question. Oscar Wilde once described 'The English country gentleman galloping after a fox' as 'the unspeakable in full pursuit of the uneatable'. We may paraphrase Wilde and describe 'the palaeoanthropologist enquiring into the origins of human language' as 'the insatiable in full pursuit of the unprovable'. Holloway, in spite of his cynicism, is enthusiastically helping to lead that pursuit.

It is the great linguist Noam Chomsky who is mainly responsible for the notion of commonality in human language, pointing out that there is a fundamental 'deep structure' to the grammar of all tongues. Apart from the characteristic sounds that issue in streams from the mouths of humans everywhere, the fundamental nature of language is in the mental ability for classifying, for formal patterning, and for relating concepts: the key trick is analysing and creating messages according to grammatical structure. Probably because no one has ever heard animals converse in a fashion implied by these rules, there arose the idea that not only is spoken language unique to humans, but that the intellectual equipment underlying it is unique too. The special set of dynamic sounds that is human language, the argument goes, is generated by mental machinery that is found only in human brains: the machinery evolved during the human career for the specific purpose of organising spoken language.

If this were true then it would draw a very neat and satisfying line between verbal humans and our less sapient non-verbal animal friends. But there are now more than a dozen chimpanzees and a bouncing young gorilla who can testify otherwise.

The University of Oklahoma Institute of Primate Studies is a few miles outside the university town of Norman. The Institute is the home of psychologist Roger Fouts, and also of 20 chimps, a group of healthy apes who regularly give visitors a noisy and enthusiastic simian welcome from their cages: they jump, sway, and swing with great agility, accompanied by a characteristic chorus of hooting. But, if you were the visitor, you'd soon notice that not all the animals' antics were purely simian. One of them would be gripping fingers of her left hand with her right; then she'd pull her left hand rapidly upwards out of the right hand's grip; and then she would point to herself. And if no one took any notice, she'd do it again. To most people the gestures would mean nothing. But to someone accomplished in American Sign Language for the Deaf (Ameslan), it would mean 'Let me out'. This is Washoe, the first non-human to learn a human 'language'.

The idea of teaching sign language to non-human primates is not new: way back in 1661 the legendary English diarist Samuel Pepys saw a baboon – or it may have been a chimpanzee – and wrote 'I do believe it already understands much English, and I am of the mind it might be taught to speak or make signs.' But Pepys's percipient suggestion was not acted upon until the middle 1960s when Allen and Beatrice Gardner started instructing Washoe in Ameslan at the University of Nevada. In 1970, when the pioneering chimp moved to Oklahoma, her vocabulary was already more than 150 signs, and still growing. Now there is a small elite of chimps learning Ameslan at the Oklahoma Institute, with Washoe the accomplished veteran.

When news of Washoe's remarkable achievements began to percolate through the primate research community, it took many people by surprise, shock even. An ape mastering a human language: preposterous! Certainly, it breached yet another barrier that was supposed to separate *Homo sapiens* from the rest of the animal kingdom. That breach widened as more and more primate laboratories explored the language capabilities of chimpanzees: some experimenters, such as David Premack in Princeton, devised a language of plastic symbols with which he conversed with Sarah; in another project, probably the most sophisticated of them all, Duane Rumbaugh of the Yerkes Regional Primate Center, in Atlanta, Georgia, developed a computer-based keyboard system that operates in an artificial symbol language he has called Yerkish – Lana is the chimp star here.

Human language is, of course, enormously complicated, so complicated that even after years of intensive effort the most sophisticated computers still fail to generate passable conversation. Little wonder then that the chimps, in their various language media, have not so far indulged in the simian equivalent of sophistry. Their conversation is at best basic. For instance, an Ameslan interchange between a *New York Times* reporter and Lucy, one of the Oklahoma chimps, went something like this: holding up a key, the reporter asked 'what's this?'. Lucy answered, 'key'. 'What's this?' he then asked, presenting Lucy with a comb. 'Comb' Lucy replied correctly. Then she said 'comb me.' He did. 'Lucy, you want to go outside?', the reporter asked. After a pause for thought Lucy declined saying, 'Outside, no. Want food; apple.' Having no apples with him, the reporter had to apologise for being unable to comply.

In the serious research projects at these various centres, the interchanges between psychologists and apes is, of course, not confined to idle chatter. Fouts, Premack, Rumbaugh and others are probing the innate language capability of their chimps, a somewhat curious pursuit to follow with animals that normally display no sign whatsoever of such language ability. Two things that all the talking chimps have done, and which is highly significant in revealing what goes on in their brains, is, first, invent new names for objects in a way that shows they can generalise, and second, show a slight but definite preference for ordering the different types of words in their conversations (a primitive simian grammar).

Generalising is something we do all the time. Chairs, for instance, come in all shapes and sizes, and yet we can instantly recognise and name such structures as basically the same thing. We have a concept of a chair in our heads, and we can slot tubular chrome creations, solid oak carvers, and plush velvet easy chairs into the same mental pigeonhole. Generalising to concepts like this is cognitively economical, and it is essential for language. Chimps can do it. For instance, Lucy calls a watermelon a *drink fruit*; Washoe refers to ducks as *water birds*, and she invented the name *rock berry* for a brazil nut when she first encountered one; Lana calls a cucumber a *banana which-is green*, and she refers to Fanta orange drink as *Coke which-is orange* (fortunately the Coca-Cola Company also make the Fanta drink, thus avoiding any embarrassing copyright problems for Lana!).

There is now no doubt that chimps have the basic mental machinery for organising a simple language. And so, it turns out, do gorillas. In uncharacteristic misjudgment Robert Yerkes said half a century ago that 'in degree of docility and good nature the gorilla is so far inferior to the chimpanzee that it is not likely to usurp the latter's place . . . in scientific laboratories.' Generally, gorillas have been assumed to be more dimwitted than their smaller simian cousins. But Koko, the young female gorilla in the charge of Stanford graduate student Penny Patterson, is busily proving otherwise. After four years' tuition in Ameslan, Koko is as accomplished as any chimpanzee and is displaying equal facility in concept formation and sentence construction.

So, even though apes can't speak, this does not mean that we can dismiss out of hand their capability for language. Arranging plastic symbols and punching arbitrarily shaped symbols on a computer keyboard may not seem much like language, and some people may feel that the hand gestures of Ameslan stretch the definition of language too far. But, if the *structures* of language are in a brain, then we have to admit that the brain possesses a language capability. More than 25 years ago the great American neurobiologist Karl Lashley said, 'I am coming more and more to the conviction that the rudiments of every human behavioural mechanism will be found far down in the evolutionary scale and also represented even in primitive activities of the nervous system.' If the argument could be put to them, Washoe, Lucy, Lana, Koko and their simian friends would agree with Lashley, and they would disagree with Chomsky's assertion that the intellectual machinery underlying language is a uniquely human evolutionary invention.

So, if our unique ability to speak is simply the product of a more general piece of intellectual equipment, what can we infer about why human languages emerged? Undoubtedly a complex of evolutionary pressures must have conspired to put words into the mouths of our ancestors, but almost certainly one of them was the advantage of being able to create better pictures in their heads. We mentioned earlier the brain's prime function of creating a perceptual world which its owner can inhabit. The ability to see commonalities between objects of the same type – classes such as trees, fruit, predators, birds etc – is a crucial step in creating conceptual order in what otherwise might be an overwhelming perceptual chaos. Presumably, chimps and other higher primates must be able to do this, and probably they go a step further and indulge in imagery to some limited extent in any case. Unless they were to reflect on past events and project them into the future – to imagine – chimps would not occasionally break off a long twig, strip off the leaves, and set out in search of a column of safari ants.

Creating and mentally manipulating images is a way of exploring your environment from the perspective of experience, and the sharper those images are in your head, the more effective you will be in exploiting that environment. Words – arbitrary sounds that name specific objects or events – are superb tools for sharpening and manipulating images in one's own head, and for invoking them in someone else's: book, storm clouds, black stallion, beautiful woman, handsome man, war – all these words may pluck images from your mind, images admittedly that differ from person to person, because of both the diversity of the world and the diversity of individual experience.

Words are powerful instruments for telling stories. They are less spectacularly successful for giving instructions. We are not suggesting that the need to sit around spinning enthralling stories to a captivated audience was a great enough evolutionary force by itself to upgrade a basic piece of mental circuitry into a fully operating language-producing machine. But who would deny that the facility to share experience – to create a genuine shared consciousness – would be evolutionarily advantageous to creatures whose unusual subsistence economy forced them into a uniquely intense social contract?

Our early *Homo* ancestors grew up and lived in bands in which individuals depended on each other for survival more acutely than is the case for any other primate. Most apes and monkeys are highly social creatures too, and if they grow up isolated from their fellows they become distinctly neurotic. But, although individuals in troops of apes and monkeys depend on each other for their psychological well-being, they are not *economically* dependent. Because of their mixed economy of gathering and hunting and subsequent food sharing, our *Homo* ancestors did need each other for economic reasons, and this tied the social and emotional knots even tighter. Sharing experience within the cooperating band through the medium of evocative language must have been an essential ingredient in the increasingly rich social mix.

The full expression of story-telling as a social cement comes, of course, with tribal myths, accounts of their people's creation. There is no human group on earth that cannot tell you how their ancestors came into being, and, with a virtually uniform desire for self-denigration, it usually involves disfavour with an all-powerful god. Nevertheless, in spite of an offended deity, each particular tribe is always regarded by its members as being special. For instance, for the Yanomamo indians, whose creation myth involves a flood, the real world is their world, and all other people are classified equally as 'foreigners'. And the !Kung's own name for themselves is Zum/wasi, meaning the real people. People who are not !Kung are not real people.

Although we shall never know, it is difficult to believe that 1470 and his fellows in their lakeside community two or so million years ago talked in reverent tones about how their ancestors had been set down by an angry god on the shore of their great lake many summers before. It is, however, not so far fetched to imagine that their 'language' was adequate for creating crude social customs and rules that were necessary to ensure basic social cohesion. Their daily lives led them into agreed divisions of activities, into journeys far afield in search of both food and materials for their basic technology, and, most important of all, into continuing and repeated inter-actions with the same individuals. Because of the ever length-ening period of infancy, a development that was essential for rearing intelligent cultural creatures, 1470 and his fellows lived in relatively stable social and economic bands over long periods of time.

Apart from humans, termites and other social insects are the only other creatures that thrive under such intense condi-tions of economic cooperation. (The social carnivores, such as African wild dogs and wolves come close, but are still in a different league.) Termites teem in millions in their awe-inspiring colonies, with social order being maintained by a combination of individuals' automatic inbuilt behaviour and stereotyped responses to the controlling chemicals (pheromones) of their own and other castes. Humans, need it be said, are not termites: apart from sucking at the breast when newly born, we have virtually no inbuilt unmodifiable behaviour patterns, and we don't respond automaton-like to the pheromones of our fellows. But, without a commitment to social order, without loyalty to the band, an individual in a gathering and hunting group would perish and cooperate activity would be impossible. Throughout human evolution, there must have been strong selective pressures for social conformity, and, un-termite-like, the structure within which people operated was provided by group culture: that culture would be impossible without a language with which to construct it.

We are not here propounding a fascist philosophy of un-bending cultural conformity for the twentieth century. We are merely trying to view the ancient emergence of human language – the instrument of culture – from the perspective of social psychology. If, as seems likely, it is true that natural selection favoured the evolution of social conformity over a period of several million years, then it would be foolish to deny that it has left a legacy in us now. If people were not so eager to conform, wars could not be so readily organised. We do not suggest that social conformity is the *cause* of war; it simply makes war easier to organise by power-motivated leaders. But, as we stress repeatedly, because humans are now such deeply cultural creatures we are by no means slaves to biological behaviour patterns. But nor are we immune from skilled manipulation of some basic impulses of human nature. This is an important topic that we will take up again later.

Chimpanzee mother
and young use sticks
for collecting ants to
eat.

Baboons, like
chimpanzees,
sometimes catch small
animals to eat. Here a
large male eats a
young Thomson's
gazelle.

Chopper

Heavy duty scraper

Chopper

Chopper

Burin

Light duty scraper

Beaked point

Discoid

Polyhedron

Tools from the Oldowan industry, excavated from Olduvai Gorge. The Oldowan industry is the earliest identifiable collection of tools, first made around two million years ago. (Scale: approx. half life-size.)

Handaxe

Cleaver

Scraper

Handaxe

Scraper

Tools from the Acheulian industry excavated from Kilombe, Kenya. Tools of this sort were a later development, being first made about 1.5 million years ago. (Scale: approx. half life-size.)

Eroded sediments in the Siwalik Hills of Pakistan where the earliest hominid, *Ramapithecus,* is found in company with other ape-like creatures.

The badland terrain of the Hadar in Ethiopia — the site of Lucy and other important fossils.

The rolling countryside near Sterkfontein, South Africa, where some of the earliest discoveries of hominids were made.

A seasonal river cuts through the sediment on the edge of the Serengeti Plain to form Olduvai Gorge.

The Yanomamo Indians of Venezuela. Hunters and horticulturalists, they are noted for their fierce behaviour.

The !Kung San of Botswana are peaceable people. Here an elder tells a story to an enthusiastic group.

A !Kung camp in a mongongo nut grove. The nuts are the !Kung's staple food.

!Kung hunters are extremely skilled at reading the slightest tracks left by prey animals.

!Kung women use sticks to dig up succulent roots.

The sounds uttered by our distant ancestors left no tangible trace in the archaeological record, so how do we look for signs of emerging language in our past? Basically, there are two ways: first, we can examine the changes in the size and shape of the hominid brain through the past few million years as this perhaps holds a clue to the neural expansion and re-organisation necessary for language machinery; and second, we can look at the material products – stone tools and ritual-istic objects – which tell us as much about what was in the minds of the individuals who made them as they do about the manipulative skills of their hands.

In most people the neural apparatus for comprehending and generating language lies in the left hemisphere: an area towards the front (known as Broca's area) coordinates the muscles of the mouth and throat that we use when we speak; and a second centre (Wernicke's area) at the side of the brain is responsible for the structure and sense of our language. Wernicke's area receives information from auditory and visual channels, and it is not a neurological accident that this impor-tant piece of language equipment lies close to a major 'associa-tion area' of the cortex, a group of nerves that integrate and compare the incoming information from all the senses. When we have something to say, Wernicke's area organises the words according to a basic grammatical form, and then sends signals to Broca's area along a nerve bundle known as the arcuate fasciculus; the circuitry in Broca's area responds by co-ordinating breathing, tension in the vocal cords, and movement of the tongue and the lips so that the right sounds come out of our mouth.

The effect of packaging all this machinery in the left hemisphere is that it is usually slightly bigger than the right. Moreover, there is something of a lump over Broca's area, and another, less pronounced, over Wernicke's area too. Searching for signs of language capability in fossil brains might therefore seem to be a matter simply of looking for the right lumps in the right place. But there are two problems. First, as Ralph Holloway points out, most hominid craniums suffer at least some distortion as they enter the fossil record. And second, biologists have very recently discovered that, for some reason that is a little puzzling, apes have a similar asymmetry in their hemispheres: one hemisphere is often bigger than the other, *and* they have a lump where Broca's area should be, not as pronounced as humans perhaps, but it is there nonetheless.

Shortly after Meave Leakey had completed the reconstruction of 1470, Ralph Holloway came to the National Museum in Nairobi to take it apart again. He wasn't being perverse: it was a necessary part of his procedure for making an endocast of the cranium. He spent a week and a half making the cast, and when it was complete one of the first things he looked for was Broca's area. There it was, unquestionably well developed. Although it is difficult to be sure it seems probable that 1470's Broca's area is more pronounced than that in apes or than similar lumps that appear in the appropriate location in australopithecine brain casts. Does this mean that *Homo habilis* (the species to which 1470 is assigned) had a more richly varied vocal output than either modern apes or its australopithecine cousins? We don't know, but the evidence is tantalisingly suggestive.

Looking at the early hominid brain as a whole, we do know that evolutionary pressures had already moulded it into a recognisably human form by at least two million years ago, and probably much earlier than that. We know too that between three million and a half million years ago *Homo*'s brain expanded substantially, whereas the australopithecines were obviously living lives that put few pressures of natural selection on their brains to get bigger. Was the initial shift into the hominid niche responsible for creating the specifically hominid brain, just as it encouraged our common ancestor, *Ramapithecus*, among other things, to walk on two feet? And did the development of an efficient gathering and hunting economy, together with its social and psychological demands (including language), drive the probable internal reorganisation and growth of our *Homo* ancestors' brains, while the ecologically less adventurous australopithecines remained more dimwitted? Again, we don't *know*, but it is a reasonable proposition. Unfortunately the proposition doesn't help us much in pinpointing very precisely when spoken words began to flower in the protohuman brain.

If Holloway's endocasts of ancient hominid brains conspire to give us no more than tantalising clues about the origins of human language, what instead can we learn from the material products of those long extinguished minds? What does it mean that close to three million years ago the tool-kits of our ancestors were crude compared with those of *Homo erectus* two million years later? And what does it tell us of the minds of our forebears that at some relatively recent point in our history they turned their manipulative talents to aesthetic as well as practical affairs? The insights we can gain are important, but to achieve them requires once again the perspective of social psychology and not just the more mundane analysis of the best way of making a living in a technologically primitive world.

For a start, what were the dynamics of technological progress in human prehistory? Given the evanescence of technologies based on wood, skin, or other plant materials, we must concentrate on stone tools: for the purposes of thinking about language origins it is not too serious a problem that stone artefacts represent only a part of our ancestors' economic activity. The earliest recognisable stone artefacts known at the moment come from the Hadar site in Ethiopia. To label them tools may seem to prostitute the word, but undoubtedly these crude flakes and choppers are the purposive products of hominid minds and hands of more than three million years ago. The tools left behind on the KBS camp site on the shores of Lake Turkana some two or more million years ago are certainly more impressive as a tool-kit than the Hadar implements, but not staggeringly so. Clearly, at this point in our history our *Homo* ancestors knew that by banging rocks together they could produce razor-sharp flakes and bigger lumps with jagged edges where the flake had been split off.

Although the gradual refinement in stone tool production that Mary Leakey documents in the material culture in the early deposits at Olduvai undoubtedly represents real progress, it is not until perhaps one and a half million years ago that a distinctly new dimension of technology arrives. This comes in the form of the beautifully tear-drop shaped handaxe that is the hallmark of the Acheulian culture. The conceptualisation involved in manufacturing these impressive symmetrical objects is of a different order from that involved in making simple core tools with sharp edges. Until around a quarter of a million years ago material culture remained modelled mainly on the Acheulian, with neither any *dramatic* technological innovation nor any great geographical variation.

It wasn't until about 100,000 years ago that technological change really began to pick up pace, with the invention of new techniques for making stone blades and an increasing complexity to the components of the tool-kit. But it still wasn't until about 40,000 years ago that the pace quickened to any-

163

thing remotely resembling the rate of change to which we have grown accustomed in our material lives today. Styles of stone-tool cultures changed relatively rapidly, and there was increasing geographical variation in cultural characteristics too. Even so, innovation was measured in millennia rather than decades. It is during these later stages, say from 50,000 years onwards, that non-utilitarian pursuits such as painting and carving started to become an important ingredient of life. Or perhaps more properly we should say that it is from this period that we have tangible and persistent evidence of such activities. As the materials of aesthetic and symbolic expression of, for instance, the Australian aborigines are mainly items such as wood, feathers, blood, ochre powder, incisions, sand drawings, songs, and dances, all of which are rapidly perishable, we will never discover any archaeological trace of our ancestors' indulgence in rituals similar to those of the Australians.

Advancement in technology throughout human prehistory is reflected in the steadily growing number of separately identifiable implements that were produced. The *range* of tool types over the period did not expand greatly, but there was an inexorable imposition of *pattern* on the types of implements in each tool-kit. Passage of time brought greater *order* to stone-tool technologies, but not an appreciably greater *diversity*.

In the previous chapter we talked of the ecological advantages of improved technology in gaining access to a greater range of energy resources. Here we will proffer an interpretation of technological change that initially may seem recklessly extreme but which we believe may be close to an essentially unprovable truth. Once again, much of the keen insight in this area of archaeology is due to Glyn Isaac.

It is quite possible that the technological element in our ancestors' economic life remained fundamentally constant throughout the period two and a half million years ago right down to half a million years ago. This is a huge tract of time, many thousand-fold longer than the whole period of the industrial revolution that has so recently transformed our material world. During those two million years our ancestors were pursuing a highly successful gathering and hunting existence in which technological demands were comparatively light. Improved social organisation, cooperation, and cohesion were the essential ingredients in the unquestioned success of this unique way of life. And we have suggested that it was these elements that provided the principal stimulus to the reorganisation and growth of our forebears' brains.

What, then, is one to make of the slow but undeniable improvement in material technology over this period, a change that, in a swirling sea of speculation, offers us virtually the only secure island of evidence about early hominid activity? The point is that there is virtually no job that can be done with the implements in the well-developed Acheulian tool-kit that cannot equally well be performed using the essentially random but effective products of enthusiastically crashing two stones together: slicing, chopping, scraping, and piercing edges, all are present in both the toolkit and the pile of rubble. The difference is that one pile of stones is more *organised* than the other. No one should imagine that anything resembling a handaxe would emerge from the fragments flying from two large stones colliding at high speed. But it is more than probable that the job to which handaxes were put could be done just as effectively by one or more of the chance products of this *laissez-faire* method of tool manufacture.

Our argument is that there would be few biological advantages in creating a tool industry that follows defined patterns when an enthusiastic but casual stone knapper can produce all the edges, points, and surfaces he might need within just a few minutes. Indeed, if laissez-faire tool manufacture were able to provide the edges and surfaces required to support the gathering and hunting economy without too much effort, it would be evolutionarily *expensive* to develop brains capable of manufacturing tools to formal patterns. So how did formalised and ordered technology emerge? The answer is that it reflected an increasingly formalised and ordered social structure, a structure that demanded the facility for a sophisticated language. As rules and customs emerged for controlling the social and practical organisational problems of operating a successful gathering and hunting economy, those rules and customs impinged on the elaboration of material technology too.

Archaeologists attach many labels to the broken and flaked stones that litter the path of human evolution: artifacts, tools, implements, tool-kits, stone tool technology, stone tool culture, all may be applied to these material products of our ancestors' hands and brains. But, with the passage of history, some terms become more appropriate than others. At one extreme, in the earliest part of the record, the word 'artifact' probably best describes the stone objects we find: they were made intentionally, but in a casual opportunistic way. As more and more formal patterning organised the artifacts into a tool-kit, then the phrase 'stone tool culture' becomes meaningful. It refers not just to the elaboration of sophisticated technology, but to the fact that technology is very much part of our ancestors' emerging social culture. Now that all there is left of those past times is the technology, it acts as a mirror reflecting the complexity and intensity of the social culture that created it.

Baldly stated, then, our thesis is that as the emerging loom of language wove increasingly organised patterns within proto-human cultural fabric, the products of stone tool manufacture also became more formalised, not because economics demanded it, but because that was the way our ancestors' minds were working: structure was stamped on social dynamics and on material technology. Admittedly, it is probably too extreme to deny that the manual dexterity required in constructing tools to a predetermined shape exerted absolutely *no* evolutionary pressure on brain development. But we suggest that it was a minor factor compared with the social and psychological pressures that required the shape of the tool to be predetermined at all. Inventiveness in the application of edges, points, and surfaces would also have added a tooth or two to the evolutionary ratchet lifting the hominid brain to its unequalled position in the animal world. Again, however, its role was almost certainly less outstanding than that of the social behavioural factors.

There are in fact remarkable commonalities between tool-making and language, commonalities in the cognitive mechanisms that operate them and in their contribution to making *Homo sapiens* the only truly cultural animal. Many animals modify their environment to some degree, by digging holes in the ground, damming rivers, and destroying trees, for example, but only humans impose their thoughts on the world around them in an entirely arbitrary and symbolic way. Both words and stone tools are inventions of the human mind; they are not simple modifications of natural objects in the way that a twig stripped of its leaves becomes an ant-dipping stick for a chimpanzee. Humans and animals share many cognitive and sensory faculties, but it is the integration of these mental elements together with the imposition of arbitrary form that makes us cultural beings.

The structure of language and the process of making tools overlap interestingly too. Both are made up of individual elements (words in language, and particular types of strokes in tool making) that have to be ordered according to a basic form (grammar in language, and the required sequence of strokes in manufacturing a particular implement in tool making). As the mental machinery required for language evolved gradually in the early hominid brain, it carried with it a 'cognitive model' for the closely related function of making tools to a particular pattern. Thus, as the cultural pressure grew for conformity in technology as well as conformity in economic and social interactions, language, which is the medium of culture, carried with it the intellectual spur for formalising technology. This is a very neat example of positive interplay in the evolution of related abilities. And we should also note that the geographical variation in technical design of stone tool cultures from about 40,000 years onwards mirrors uncannily the geographical patchwork of different languages with which we are familiar today.

One has to reflect only for a moment on any normal conversation between two people, or perhaps imagine the performance of a powerful orator, to glimpse yet another element of human dialogue: non-verbal communication, particularly gestures.

Facial expression, inflection in the voice, gestures of the hands and arms – all play an enormous part in human communication. This channel of communication is, of course, the only one open to the non-verbal animal world, and during recent years biologists have discovered just how underrated it has been by the scientific community. Remembering chimps' and gorillas' mastery of American Sign Language, albeit taught by human instructors, one inevitably ponders on the possible role of elaborate gestures in hominids' early excursions into more sophisticated communication and symbolising. If modern apes can learn symbolic signs and use them in a structured way, it doesn't seem unreasonable that Pliocene hominids (between five and two million years ago) with their more humanly organised brains should have been able to invent their own sign language.

Gesture language, combined with expressive vocalisations and facial movements, may well have been hominid evolution's 'first attempt' at breaking through the communications barrier, only to be supplanted later by a vocal language with its infinitely greater capacity for handling information. Possibly the development of more expressive gestures and vocalisations (leading to spoken language) began simultaneously, with the spoken word proving the more powerful. Gordon Hewes, an anthropologist at the University of Colorado, is a particular proponent of hominid gesture language, and he points out that the increasing manipulative skills required for manufacturing the more complex stone tools may be not unrelated to manual skills associated with complex signing.

It would be difficult to refute the idea that meaningful gestures have been an important component of human communication for a very long time. We all gesture to some degree when we speak – particularly so when we are stuck for words! And the universality of many types of gestures and facial expressions throughout the world argues that manual signing as a means of communication is deeply planted in the human brain. What we now experience in ourselves and see in others must be the remnants of an ancient means of communication of some sort, but whether it was a forerunner of spoken language, or a concomitant of it, is impossible to say. However, we take the parsimonious view that the cognitive leap to the beginnings of a spoken language was not so great as to *demand* the interpolation of a system of gestures.

When we skim through the pages of the archaeological record, the evidence not only grows steadily more profuse as we move towards the present, it also begins to include nonutilitarian objects: carved figures, pendants, wafer-thin flint 'laurel' blades, and paintings. These products of an aesthetic mind, conceived in a context of symbolism, begin to be a major aspect of archaeological sites only from about 30,000 years ago. But the culturally inspired motivation that was clearly gaining a considerable momentum by this time must have had its first tangible expression long before. The piece of sharpened ochre used by the inhabitants of the springtime home at Terra Amata 300,000 years ago was probably used for some symbolic purpose, but as the lines that were drawn with it no longer exist we are left only with empty speculation.

There is, however, a single relic of our ancestors' aesthetic symbolic sense from the same period as Terra Amata, and it came from a place known as Pech de l'Azé in France. There, 300,000 years ago, a time when our ancestors were in transition between *Homo erectus* and primitive *Homo sapiens*, someone sat with an ox rib and carved on it a series of festooned double arches. Why, we don't know, but the pattern is exactly the same as that inscribed commonly throughout the period 40,000 to 15,000 years ago. After the carved ox rib there is a gaping void in the archaeological record until more carved objects appear, such as a shaped mammoth molar tooth worked on by a prehistoric craftsman 50,000 years ago at a place called Tata in Hungary. We don't imagine that our ancestors gave up converting their symbolism into tangible form during that enormous gap: there must be countless examples of ancient carving just waiting to be found.

One of the earliest examples of 'art' is a two-inch-long statue of a horse carved from mammoth ivory more than 32,000 years ago. It comes from a cave at Vogelherd in south Germany and it is the oldest of several statuettes from the same site. Produced with an artistic skill equal to that of the high point of cave painting in Europe and Africa 20,000 years later, the horse had been carried around in a pouch and handled over a period of several years: the eye, ear, nose, mouth, and mane, which had clearly been carefully carved, were worn down by prolonged handling. In fact, all the Vogelherd statuettes revealed the same tell-tale signs of repeated use: they don't appear to have been ornaments of self-decoration, but they may have been personal 'good luck' charms constantly fingered absent-mindedly during daily chores; or perhaps they figured in repeated social ceremonies.

Whatever the specific use of these carvings and paintings, they speak to us of a truly human spirit that we all recognise in ourselves. We may not be the only animal who intentionally makes tools, but are the only creature who sets out to manu-facture strictly non-utilitarian objects. It is the mark of a fully maturing culture.

What has this to do with language? Just as the production of a formalised stone tool-kit is indicative of linguistic capability, so too is the elaboration of strictly symbolic creations. But, although it is just about *conceivable* to imagine complex stone tool cultures being manufactured by non-verbal creatures, it is *impossible* that abstract symbolism, such as we see elaborated during the past 30,000 years, could arise in a speechless animal. Without words with which to name them, a statuette of a horse, a rock painting, or a nation's flag, could never exist: they would be meaningless.

The rapid acceleration in the pace of technical and aesthetic progress that started 30,000 years ago was the signal that some kind of critical point had been passed in the accumulation of cultural experience. The physical evolution of humankind was slow and steady, as is the nature of such biological change: fully modern *Homo sapiens* emerged around 50,000 years ago, perhaps involving a final refinement of spoken language. But when the slow fuse of culture, which had been smouldering gently for perhaps two million years, reached its critical point it ignited an explosion in human progress that roars on with ever-increasing force.

As cultural creatures we can impose our will on the environment and on ourselves, depending on the social rules we choose to make and follow. Culture allows *Homo sapiens* to lead lives of a diversity within the species that cannot be achieved by any other animal. Each one of us is born with the potential to live any of countless different lives, but we live just one, the one shaped by the rules of the culture in which we grow up. In viewing humanity we should not imagine ourselves as animals fortunate enough to have been endowed with the added advantage of culture, a cloak of human respectability thrown around a mute animal nature. Without culture we would be neither human nor animal. Clifford Geertz, a brilliant American anthropologist, put it this way: 'A cultureless human being would probably turn out to be not an intrinsically talented though unfulfilled ape, but a wholly mindless and consequently unworkable monstrosity. Like the cabbage it so much resembles, the *Homo sapiens* brain, having arisen within the framework of human culture, would not be viable outside of it.'

CHAPTER 10

Sex, and the need for women's liberation

'In the beginning,' wrote Freud, 'was the Deed.' The 'Deed' was the great man's view of how a primitive ape-like creature was transformed into a recognisable human ancestor destined to suffer the psychological anguish that keeps modern analysts in business. In the beginning, he said, humans lived in family hordes dominated by a single masterful male. The sons, when they became mature enough to pose a threat of sexual competition to the male who sired them, were ejected from the horde. The ejected brothers eyed with envy the females which their father monopolised, and they formed themselves into disgruntled bands awaiting their opportunity. 'One day.' Freud speculated, 'the brothers who had been driven out came together, killed and devoured their father and so made an end to the patriarchal horde. Some cultural advance, perhaps command over some new weapon, had given them a sense of superior strength.'

This evolutionary breakthrough was not without its psychological scars, Freud tells us, for, although the brothers hated their father, they felt guilty about what they had done. To mark their guilt they invented totemic prohibition and incest taboos: the totem animal that must never be killed in hunting, except on special occasions when the brothers restated their solidarity; and the incest taboo meant they would eschew any sexual contact with their father's women, who indeed had been the cause of the trouble in the first place. This is the Freudian view of human psychology, and it is from this that arises the fundamental feature of our psyche, the Oedipus complex.

Freud, and others such as Malinowski and Westermarck, were trying to unravel the contribution of *imposed* cultural rules and *natural* primate tendencies to the nature of human behaviour. Remembering what Clifford Geertz says about the inextricability of the two factors in the quintessentially cultural *Homo sapiens*, one must tread very cautiously in this treacherous territory. As with Freud, our preoccuaption here is with sex, and for good reason because, apart from our unusual protruding nose with downward-pointing nostrils, the outrageous sexuality of humans is one of the few characteristics that truly sets us apart from our primate cousins. Notwithstanding the unlikely – and more than a little impractical – passion that burned in the outsized breast of King Kong for his voluptuous young lady, *Homo sapiens* is the only primate for whom sex *as an activity* is such a consuming preoccupation. Assuming that the influences of evolution endowed us with our extreme sexuality because it was biologically advantageous rather than simply as an added element of social complexity and potential disruption, we shall try to explore those putative advantages.

Although individual human sexuality is a unique phenomenon in the animal kingdom, styles of mating – from polyandry, through monogamy, to polygyny – are an important facet in the social structure of all animal communities. Biologists are only now beginning to understand the factors that are important in favouring one mating style rather than another in a particular species. The whole business of sex is turning out to be much more complicated than people imagined just a few years ago, but it clearly involves a complex interaction between basic genetic imperatives and the way the species makes a living.

The question we want to ask when we are contemplating human origins is, can we peer back into the human fossil record and reconstruct a picture of this important sphere of social interaction in the early hominids? As British anatomist Alan Walker said at a conference in California at the beginning of 1977, 'If all we can get from our work is a list of hominid species in space and time, then we should give up: we want to know how they lived, how they survived together, and what their sex life was like.' It is possible now to explore the sex lives of our ancestors without resorting to mere fantasy, and that is our intention in this chapter. We also plan to reflect on the more thorny problem of what a view of our past might imply for an interpretation of modern society, both in terms of individual sexual behaviour and social structure. Particularly, why is there a virtually universal domination of men over women in the spheres of politics, economics, and social status?

As we said in an earlier chapter, the rules of natural selection demand that the aim of individuals engaged in the game of evolution is to produce as many surviving offspring as possible. If parents, through some fortunate genetic quirk (a mutation), give rise to young that are particularly suited to thriving in the prevailing environment, or perhaps to exploring successfully a slightly different one, then it will be this genetic stock that will predominate in the future: individuals whose genetic blueprint makes them well adapted to their ecological niche will thrive, while those whose genes make them less able to make a living will eventually die out. Through the long eons of biological time there is a slow but steady drift in animals' genetic makeup so that the behavioural effects of new or modified genes are 'tried out' against prevailing conditions: if the new genes are advantageous they survive and thrive, if not they disappear. This, very simply, is how new species arise while others become extinct: in the competition for exploiting resources in the environment, some animals succeed, some fail.

The revolution in biological science that has been stirring quietly for some years but which has only very recently gathered any real momentum concerns the focus of evolutionary pressures: biologists are now beginning to realise that, because of the laws of natural selection, it is not just different species that are in competition, but individuals within a species compete with each other too. Natural selection works at the level of individuals, not of populations. As we said, if a genetic mutation in an individual is particularly advantageous, then it is that individual's genes that will increasingly form the genetic stock of future generations of its particular species.

An individual will behave in a way that is beneficial to itself: it won't out of *pure* altruism give a helping hand to the genetic success of a competitor. The altruism we talked of earlier was between related individuals who therefore have some genes in common, or between individuals who, because of their social structure, are sure to repay the favour. In these situations, apparently selfless behaviour is in fact fundamentally selfish. In all of this it must be stressed that none of the behaviour – the competition or the various forms of altruism – is conscious or intentional: it is simply a form of behaviour that has evolved because it is what biologists call adaptive – it enables the individual to survive and reproduce more successfully.

173

What has all this to do with sex? The point is that in the race to produce as many surviving offspring as possible, males and females, strange as it may seem, are in competition with each other. Now, if sexual reproduction for an animal simply means depositing eggs and sperm in the same place at the same time – as frogs do – then there isn't too much of a problem. Even so, anyone who has ever seen the gelatinous mass that a female frog lays can readily acknowledge that her contribution to the next generation in terms of materials and energy is vastly more substantial than the filmy offering of microscopic sperm provided by the male. Both eggs and sperm are necessary, of course, for that is the nature of sex, but the biological cost to the female far outstrips that to the male: her investment in the next generation is greater than the male's, and this is a pattern repeated throughout the animal world.

It so happens that when frogs have performed their conjugal activities, they leave the fertilised eggs to mature and develop with no further parental interest. As soon as any parental care is required in rearing young, however, there the battle of the sexes begins. In terms of natural selection, both males and females would prefer to deposit their sex cells, have them fertilised, and then go off and do the same again elsewhere so as to produce as many offspring as possible. But, if the maturing embryo and young need nurturing, who is going to be responsible for doing it? If infant-rearing can be coped with by just one adult, then the question seems to be settled by who can escape from the conjugal scene first, for that parent is free to go and meet another partner while the one who is left behind will have to take on the role of looking after the offspring.

In the animal world of single-parent families, it is nearly always the female who is left holding the baby, because in the race to move on to other partners they usually come second. The major exceptions are in the aquatic world of fish, and the reason is not difficult to understand. Like frogs, most fish release their sex cells into the water where fertilisation occurs. Now, because the male's sperm are so light they are in danger of drifting away if he deposits them before the much heavier eggs arrive. So, in the interests of not wasting his sperm, he has to wait until his mate has made her contribution before he can follow suit. Because of this sequence of events, the female can seize her chance and desert her mate, leaving him whatever work there is to do in ensuring the safety and satisfactory nutrition of the brood. The male is caught in what Robert Trivers calls a 'cruel bind', for the male could decide to desert too, but then there would be no offspring at all. It is therefore not surprising that natural selection has favoured the evolution of *paternal* devotion in many fish.

The downfall of females came when animals moved on to *terra firma*. In the comparatively hostile conditions of dry land, animals could not leave naked fertilised sex cells exposed to the air, otherwise they would rapidly dry out and perish. So eggs evolved as a response to the new environment. Once again the female's investment in the embryo is enormous compared with the male's because she manufactures the super-structure of the egg and then has to pack it with nutrients to sustain the growing embryo. In birds a female's egg burden is sometimes as much as a quarter of her body weight, a very expensive investment indeed. Moreover, because the female manufactures the egg inside her body after fertilisation, this gives male birds the opportunity to desert and sow their seed elsewhere. But, as it turns out, most of them don't: around 90 per cent of bird species are monogamous for one season at least. And the reason is that the degree of effort required to rear the young is too great for one adult to handle: prolonged incubation and the subsequent non-stop business of collecting food for the ravenous brood needs the cooperation of two adults. So, if a male were to desert his fertilised female, the chances are that his potential offspring would perish: this is not biologically sensible for the male, so he stays and helps with the family duties.

The state of avian monogamy is principally responsible for the spectacular plumage that splashes swift, bright colours across our countryside, particularly in the mating season: the males are competing with each other for the favours of the females. Because the state of food resources might mean that not all the females are able to breed, a male must make even greater efforts to catch the eye of a female if he is to have any chance of fathering offspring. And he wants to be sure of acquiring the most outstanding mate possible rather than making do with a scrawny weakling that would not do justice to *his* genetic investment. His brilliant coat and elaborate courtship are designed to impress on the female his desirability as a mate. The female, because in any event she will be left a clutch of maturing eggs, has to be convinced that the male is not only suitable as a father (to produce fine healthy off-spring) but also as a husband (to remain with her to rear the brood).

The choice of mates among these birds rests with the females, as it usually does in the animal world, and the males compete enthusiastically to outshine each other. In many species of monogamous birds, females decline their conjugal favours until the solicitous potential mates have constructed a nest that is judged suitable for housing their future offspring. This kind of activity to some extent equalises the parental investment in the yet-to-be-fertilised offspring: the female manufactures an egg and the male builds a nest. Courtship is critical for the male, not simply to win for himself a potential mother for his offspring, but also to satisfy himself that she is a fine, healthy specimen who will be able to contribute effectively to rearing the brood. But more than that, males have to ensure that they are not being tricked into settling down with a partner who has been fertilised by another male who has since departed. This would be in the interests of the deserted female because she would then have a partner to help her rear the young, but it would be disastrous for the cuckolded male: he would be investing his effort in the care of another male's genes. A long courtship before copulation may reveal whether the female is already pregnant. If she is, the male should go elsewhere for the sake of promulgating his own genes.

The uncertainty of parenthood is a constant genetic 'threat' to males. Because females lay their eggs (or give birth to offspring), they can be certain that they are nurturing their genes when they feed the young. A male, even if he copulates with his partner at the appropriate time, can never be *sure* that he's not been cuckolded. It is in the males' genetic interests to cuckold other males, and to avoid being cuckolded themselves.

Polyandry (one female mating with many males) is rare in the animal world, and where it does occur it is most commonly in birds. Even so, less than one per cent of birds are poly-androus: ecological circumstances simply don't favour it. Polygyny, the opposite of polyandry, does happen in some birds, and it depends not only on one parent being able to raise the brood, but also on the ability of one male to command significant environmental resources. One rather extreme example is the orange-rumped honeyguide which lives in southern Asia. The bird leads honey-loving animals to the bees' nests which are usually perched on the exposed faces of cliffs. The animal plunders the bees' nest, and the honeyguide then feeds on the debris. A male that can take control of a wrecked nest in the face of competition from his fellows has a sexual bonanza because the courting females come to the nest to copulate. One bird-watcher saw a lucky male copulate 46 times with at least 18 different females, whereas neighbouring males with no bees' nest had virtually no sexual success.

The benefits of polygamy, in terms of contributing one's genes to the next generation, are enormous because the individual can then have a large number of offspring with relatively little parental investment: it is the natural state to aim at, both for males and females, if the demands of child care and the distribution of ecological resources permit it. But, because of basic biology, males have many more opportunities for polygamy than do females. For a start, even if the female of a species can so arrange matters that she mates with several males, each of which then rears the resulting offspring, the exercise is still costly to her: she has to produce many batches of large sex cells each equipped with its own survival rations. For a male, polygamy simply means depositing packets of extremely cheap sperm into as many females as he can persuade to copulate with him. Because of the way the biological dice are loaded, polyandry is rare, monogamy is relatively rare (except in birds), whereas polygyny is common. In mammals, including the higher primates, some degree of polygyny is virtually the rule. Among primates only the acrobatic gibbons, and their equally agile relatives the siamang, are monogamous. (For those who view the natural order of things as having been created by a divine hand, this sexual imbalance (injustice?) may be enough to persuade them that god was male!)

Naturally, in a polygynous social system some males are going to be sexual failures: not every one can win in the battle to monopolise the sexual favours of the females. In this sort of situation it may appear to be biologically adaptive for females to give birth to female rather than male infants. At least this way the parent is assured that her genes will pass through into subsequent generations: the daughter is bound to mate and bear offspring with whichever male is most sexually dominant; a socially incompetent son may never have a productive sex life, leaving no offspring at all. But, it seems that the potential genetic benefits of having a sexually successful son are so great that a daughter-only strategy is not pursued to any degree within primate populations. The sex ratio of births is in fact roughly equal.

There *is*, however, a slight imbalance in who has sons or daughters in the primate world: the most dominant females tend to have more sons than daughters, whereas the less socially successful mothers usually have more daughters than sons. The *mechanism* by which this happens is that male foetuses are more vulnerable to stress than female foetuses. And, because mothers low down in the hierarchy usually suffer more social harassment than those higher up, some of the male foetuses abort. The genetic *rationale* for this is that in higher primates the social status of sons usually reflects pretty accurately the social standing of their mothers. So, if you are a female at the bottom of the social pile, then it is probably better to have daughters as the chances of your sons becoming socially dominant are limited. The reverse is true if you are a mother with a high social position: your sons are likely to be socially influential and therefore a big hit with the females.

The ideas about a genetic imperative shaping social structure and mating systems received some persuasive, if gruesome, support in recent years with the discovery of systematic infanticide in Hanuman langurs. These sleek leaf-eating monkeys are very successful inhabitants of the Indian subcontinent. They live in troops of about 25 females and young with usually only one fully grown male: he is the harem owner. Between 1971 and 1975 American primatologist Sarah Blaffer Hrdy studied the social politics of life in five troops of langurs around Mount Abu, an area south of Jodhpur in northwest India. During those years Hrdy observed scientifically what many anecdotal reports had suggested: that when a male manages to oust an incumbent harem owner from a troop, the newcomer kills all the young infants. This is not sheer bloody-mindedness: it is biologically very sensible. The reason is, as harem owner, the male does not want to be in charge of females who are busy caring for the offspring of another male (those of the previous harem owner). So, when a new male takes over a group of females he dispatches the young, and then inseminates all the females so as to raise *his* brood, that is, of course, if he can maintain control long enough to enable the new batch of offspring to mature beyond vulnerable infancy.

Hanuman langurs are not the only animals to obey this particular genetic imperative: lions, Barbary apes, rhesus monkeys, gorillas, red-tailed monkeys, howler monkeys – these and many other polygynous harem owners do the same thing.

The trap of high maternal investment in offspring becomes particularly fierce in mammals, because, not only do females provide a bigger sex cell, they also nurture the growing embryo inside their body for a relatively long period. And the trap continues after the infants are born too, with the need to provide milk. Because nature has not seen fit to provide males with nipples capable of producing milk, male mammals have a remarkably easy time on the domestic scene, with their paternal investment in ensuring the future well-being of their genes often being limited simply to providing sperm to the female in a swift and emotionally indifferent copulation. Among higher primates the bounds of fatherhood frequently extend only minimally into toleration of playful infants. Even gibbon fathers rarely indulge in much direct child care: they spend a lot of their time defending the couple's territory. Siamang fathers do, however, frequently carry their offspring.

When we look at the two apes most closely related to humans, gorillas and chimpanzees, it is fascinating to discover how very different their sex lives are. Gorillas live in groups of perhaps eight or 10 individuals dominated by a large silverback male. The young bachelors wander around either singly or in all-male groups of their own, awaiting their chance to own some females (this of course is very much the sort of model that Freud contemplated). The silverback achieves his position because of his magnificent stature and overt masculinity, features that attract the females to him and make his would-be successors think very carefully before trying to oust him. The silverback is usually almost twice as hefty as his females, a degree of sexual dimorphism that is very characteristic of this type of polygyny. (Males and females in monogamous gibbons and siamangs, incidentally, are the same size as each other.)

By contrast, chimps live in more mixed troops, with the core of social life being the bond between mothers and their infants. From the many thousands of hours of observations on chimp social life that have now been logged up, it is clear that siblings develop a strong affection and responsibility for each other too. In normal troop life groups of mothers and offspring usually forage together while bands of males wander in their search for food over a rather wider territory. There is no concept of fatherhood in chimp society as the males are fully promiscuous. When a female comes into oestrous, which she does once every year or so, she advertises the fact by developing a bright red swelling around her genitals. The males respond to this unmistakable signal by becoming very solicitous of her, and the female uses her instant popularity with the men as a means of comparing their various simian talents: she exercises some choice in her sexual mate, but she by no means limits her brief sexuality to one copulation: males often queue to take their turn, and the female usually objects only if the would-be mate is a son or a brother (Freud would have been intrigued to see such a powerful incest 'taboo' in such a lowly creature!).

It is perhaps intriguing to reflect that the degree of moral indignation over incest in human society is greatest when the act is between mother and son, somewhat less when it involves brother and sister, and less still when a father makes love with his daughter. This parallels exactly what one sees in chimp society, there being no natural barrier between father/ daughter incest at all. This last form of incest is limited, however, because young females usually leave the troop in which they were born as they grow to maturity.

Compared with olive baboon males, for instance, who use their more rigidly defined social status to gain sexual access to more than their fair share of receptive females, chimp males in their promiscuity are relatively tolerant of each other. They do, however, indulge in a more subtle brand of sexual competition by ejaculating huge amounts of sperm from their enormous testes: he who injects most sperm has the greatest chance of fathering the offspring!

In spite of their good-humoured promiscuity, chimp males are substantially bigger than their females (by about 20 per cent, the same figure as for modern *Homo sapiens*), thus betraying in their frames the social competition one sees in some aspects of their behaviour. Olive baboon males, by contrast, are twice as big as females, a massive sexual dimorphism reflecting their intensely competitive social life and also perhaps their more necessary role of occasionally defending the troop against predators when fleeing to the trees is impossible.

What can we say about the sex lives of early *Homo*? Sexuality does not become fossilised; nor, unfortunately, do testes, so we don't know whether or not our ancestors followed the chimps' example of subtle competition. On balance, however, it is much more likely that early *Homo*'s sex life was more like the chimp's than the baboon's. We say this because it is easier to envisage the more relaxed chimp-like society shifting its subsistence methods from essentially solitary feeding to a food-sharing economy than it is to imagine the same thing happening in the tension-filled, more rigidly hierarchical, baboon social system.

The one piece of solid evidence that does to some degree relate to sexuality is the difference in size between males and females of primitive *Homo*. At the moment there simply aren't enough fossils to judge with any confidence whether sexual dimorphism was substantial or not in *Homo habilis*. In a few years, however, this will become possible. If the difference in size between male and female *Homo habilis* turns out to be substantial, then this would signify sharp social competition between the males. A *substantial* dimorphism seems unlikely, but the difference between males and females two to three million years ago is unlikely to have been *less* than the 20 per cent that exists now. One may ponder on what this degree of human sexual dimorphism implies for the relationship between the sexes in the past, and in the present for that matter. Men may have been 20 per cent bulkier than women for ecological rather than social reasons. But at the moment there is no sound reason for denying the possibility that in our recent past men vied with each other to a limited degree within the framework of a cooperating community, and the more socially and economically successful males may well have commanded greater attention and respect from the females. The competitive edge cannot have been very sharp, however, because of the essentially interdependent nature of the economy. This inescapable biological conflict may have encouraged more subtle cerebral forms of competition rather than overt confrontation.

The transition from a basic ape-like creature into a proto-human would not have decreased the early hominid female's investment in her offspring. She would still carry the foetus, of course, and, indeed, because of the lengthening period of infancy, the subsequent commitment to child care would be increased. In the absence of milk-producing nipples, males could take no part in feeding the offspring. The simian bond between mother and child would have remained the core of social life throughout much of the human career; some people argue that it remains the prime bond even now.

The major factor responsible for the evolution of humankind was the development of the mixed economy based on sharing food. This must have affected the hominid sexual system, and it might be responsible indirectly for our uniquely human sexuality. Although a chimpanzee mother can provide for herself and her growing infant throughout its childhood because the food supply is readily at hand, an individual who is part of a food-sharing economy depends, by definition, on others for at least part of her diet. So, just as a mother relies on the male scavenger/hunters for meat, so too does the infant. The mother would have been able to take her baby with her on vegetable-gathering excursions using some kind of sling, as do modern gatherer-hunters, but the dependence on males would still be inescapable. The change to a food-sharing economy therefore took human society a step nearer to the lifestyle of many birds, where two parents must work to feed the offspring. The parallel cannot be taken too far, of course, because, instead of existing in isolated nuclear families, the essence of the human economy is that every individual depends on every other for subsistence; this is not so for any other creature, apart from social insects and, to some extent, social carnivores.

The question that has to be answered is how did the male's inevitably greater commitment to child-rearing manifest itself? Did monogamy rapidly become the norm, creating nuclear families within a small gathering community? Or did males simply provide meat as a cooperative venture for the whole of the band – males, females, and infants? Throughout human history individuals in gathering and hunting bands must have had their needs provided for on many occasions for no better reason than that they were members of the band. But the basic structure within which modern gatherer-hunters operate is one of kinship, and a kinship that involves an acknowledged father as well as a mother. There is a very powerful impression that acknowledged fatherhood stretches a long way back into the career of *Homo sapiens*: powerful bonds arose between parents.

Part of that impression comes from the fact that women are so sexy. All male primates are ready for sexual encounters at any time, and human males have simply followed the tradition. But female non-human primates lead very subdued sex lives, becoming sexually receptive at intervals that are frequently separated by years rather than months. Human females, by contrast, are interested in sex all the time. Moreover, there is more to sex than procreation, whatever the church may have to say on the matter. Non-human primate females become sexually receptive when it is biologically appropriate for them to conceive, and the event is marked by a brief copulatory encounter (usually less than half a minute) with one or more males. Human sexual encounters between acknowledged mates may be as frequent as every day, they are long lasting, they occur whether or not the female is already pregnant, and they involve the phenomenon that has already filled the pages of countless psychology books and magazines: the female orgasm, something quite unique in the animal world. As a biological response to female sexuality, human males have evolved a penis that is larger than any other primate, including the gorilla whose body bulk is almost three times that of a man's. As everyone is aware, human sexual encounters, unlike those between any other animal, involve much more than simple unemotive copulation. Some writers have even suggested that one of the most powerful evolutionary forces that caused us to become naked was the enhancement of pleasurable tactile stimulation during sex. There is no way in which this suggestion can be tested scientifically.

Female sexuality is a matter of considerable interest to men and women. From a biological point of view one may interpret this either as a mechanism for attracting the attention of several males, from which she can choose the most desirable; or it can be seen as a way of maintaining the domestic co-operation of a male long enough to provide for the economic demands of a growing infant. Male attentions are rarely in short supply, even if the males themselves are: one male can inseminate thousands of females if necessary. Most probably, then, heightened human sexuality evolved as emotional cement to an economic contract in which the product is children. In other words, sex became sexy for humans – particularly for females – as an essential ingredient in the uniquely inter-dependent child-rearing bond of *Homo sapiens*. If our ancestors had not invented the food-sharing economy of gathering and hunting around three or so million years ago, we would be neither as intelligent as we are today nor so interested in each other's sexuality.

In the choice of mates among early *Homo* bands, the same genetic imperatives that operated in the rest of the animal kingdom would inevitably have made themselves felt here too. A male would want a woman who could provide strong children and who would be able to look after them; he would be interested in her skill as a mother and as a gatherer of plant foods. And the woman would choose a man who was strong, socially adept, successful in the quest for meat, brave in the face of danger, and good with children as well as being committed to providing for them; in other words, she would want a male who was both father and husband. Although it would only rarely benefit a woman's genetic contribution to future generations if she were to seek sex outside her economic contract (the number of children she could have is limited), a male would gave everything to gain genetically by illicitly fathering children outside his 'marriage'. Natural selection should favour the evolution of Don Juans, and also the ability of females to detect the untrustworthiness hidden behind a Don Juan's veil of seductive charm. This is the rule of biology.

In a community in which child-rearing depends heavily on the contribution of two adults, the female has to be very careful in choosing her mate because she must be sure that he will stay with her. As for the male, he must bear the age-old uncertainty of the fatherhood of his mate's children: are they his or has he been cuckolded?

The deep emotional attachments between men and women are almost certainly the product of the biological need for the commitment of another adult in the care of the child, and, as we've said, sexuality is a way of strengthening that bond – it makes it deeply rewarding. But this does not necessarily imply that monogamy is the natural state of humankind. As with all biological systems, the human reproductive unit is responsive to factors in the environment. Where making a living is sufficiently easy for a man to be able to support more than one woman, we can expect polygyny. Again, biology would demand it. Polygyny is more genetically rewarding than illicit sex because the male can be more sure that the women are having his children. However, it may well be in the female's interests to monopolise the attentions of one male, whatever the ecological circumstances, in which case she may use such weapons as jealousy and the demand for a long courtship before entering into a sexual relationship to restrict illicit sex and keep polygyny at a low level.

In contemporary life monogamy is the norm only where society enforces it through laws. Where there are no such laws there is exactly the kind of spectrum one would expect: because of the economic aspects of human activity monogamy is commonest of all, but it is closely followed by polygyny; polyandry is rare and it occurs in unusual circumstances.

In contemplating behavioural aspects of human origins one must consider seriously the dictates of the various genetic imperatives we have mentioned. Certainly, they must have played their part in shaping in an unconscious way the style of our ancestors' sexual relationships. How important they are in today's world is quite another matter, and a controversial one at that. For a start, we no longer live in small gathering and hunting bands; instead we mingle with countless thousands of potential sexual partners in an intense urban atmosphere. This must perturb any biological programme, however basic, that may have become embedded in our brains.

Notwithstanding this complication, how are we to view the fundamental genetic imperatives that tell us that: women should be more fussy than men in choosing their sexual partner and that they should look for someone who will be socially and economically successful; men are more naturally adulterous than women; female prostitution should be more common than male prostitution; men are likely to be more severe with adulterous wives than wives will be with husbands who seek illicit sex elsewhere; that coupling of older men with young women should be more common than the reverse; that a maternal grandmother is more certain about her genetic investment in her daughter's children than is a paternal grandmother and will therefore be more solicitous in offering help in caring for the infants; that under favourable economic circumstances men will seek to be polygamous, and that such circumstances are likely to be more common than those conducive to polyandry; that the significant difference in body size between men and women implies that in our recent history men competed with each other for more than their fair share of desirable women; and that monogamy is not the *natural* state of humankind?

These are some of the more important facets of human behaviour one would expect to see if *Homo sapiens* were ruled solely by the laws of genetics. The fact is, of course, that we are not animate automatons following the dictates of our genes: we were conceived in the animal world, but our minds came to maturity under the influence of a self-generating culture. But, although culturally based variations mean that there is no such thing as uniform behaviour throughout all human communities, it is equally true that we *do* see a reflection of that list in modern society. It is an ugly reflection because it implies rank male chauvinism in the world of sexual relationships. The modern world *is* pervaded by male sexual dominance, not universally of course, but in the great majority of societies. Is this because even highly cultural *Homo sapiens* cannot fully escape the commands of genetic imperatives? Or is it, as some people argue, the culmination of a gigantic male conspiracy perpetuated by the artificial norms of culture?

No one can in all honesty answer this question in complete scientific confidence. One's answer will depend largely on one's view of the world. Before we offer our view we should reflect on the social organisation of the contemporary non-agricultural people in the hope that there might be valuable clues for swaying judgment in one direction or the other.

Throughout all contemporary gatherer-hunters there is a consistent division of labour between men and women: hunting meat is a male pursuit, whereas gathering plant foods is the responsibility of the women. The division isn't absolute, of course, as men don't pass by a nut grove without gathering some, and they would not ignore signs of ripe tubers if they had time to dig them up. More frequently, though, the men tell their womenfolk about newly discovered patches of nuts, tubers, etc, so that they can be exploited later by serious gathering expeditions. And women don't turn their backs on an opportunity to catch a small animal if one chances their way. Moreover, on communal net hunts such as those organised by the Mbuti pygmies of the Congo basin and the Bihor of India, men and women work together to secure the game. But, in spite of the 'grey areas', it is possible to draw a line between the *main* economic contribution of men and women in these societies.

Why the pursuit of game is principally a male prerogative is not easy to say. We can be sure that it is not purely a male conspiracy, however, because, although in baboons and chimps it is males who are the hunters, in lions it is the females who do most of the killing. As with female baboons and chimps, gatherer-hunter mothers are usually involved in some stage of child care throughout their adult lives. The burden of carrying infants on food-gathering expeditions is great (!Kung women walk more than 1500 miles a year with a suckling infant on their backs), but they are clearly equal to the task. But, although collecting plant foods is not without hazards from potential predators, the dangers of hunting are greater. The loss of the child to the hungry jaws of a carnivorous cat would be a serious blow to a woman's reproductive career: to put the child at risk would therefore not be biologically sensible, either for the woman or her mate. What is more, the woman herself is valuable as a future child-bearer: in terms of the reproductive potential of a group, the loss of a man's life – either in hunting or in war – is less of a problem than the loss of a woman's. There would also be the not insubstantial problem of persuading a two-year-old not to make infantile utterings during critical silent stages of stalking prey!

For these and other reasons, women only rarely hunt, and this is an economic strategy that has immense social and political repercussions. As we said in an earlier chapter, meat is more than body fuel: although it is relished as a specially enjoyed food, it is also a prestigious item of exchange. Plant foods, the insurance policy of the vast majority of gathering and hunting people, are shared within the limits of the immediate family, whereas hunters may distribute their game throughout the whole band. Why flesh should be regarded as a special form of food while plant foods on which most people's lives depend are thought of as mundane is something of a mystery. Most gatherer-hunters are particularly excited when meat comes into camp, because, as they very readily tell you, 'it tastes good'. But perhaps the reason runs deeper, having something to do with the spirit of life meat once had, and which plants lack.

Significantly, chimps also consume their occasional meaty meals in a way quite different from their normal eating habits. An observer at the Gombe Stream Reserve once saw a group of chimps spend the whole day engaged in the business of sharing and eating the body of an infant monkey that could not have weighed more than a few pounds. This kind of disproportionate attention to meat eating is usual among our simian cousins. Individuals lucky enough to have a morsel of meat spend many hours eating it, often savouring it with a handful of leaves. And chimps who have no meat are prepared to beg patiently for as long as four hours, and with no guarantee of success. Why?

Perhaps the mystique and ritual that frequently surrounds killing and eating game in non-agricultural people is a cultural embellishment of a more deeply rooted biological instinct of the sort we see in chimps? But whatever it was that elevated meat to its prestigious position in gathering and hunting societies, it incidentally placed social and political power into the hands of men rather than women. As the collectors of meat in the mixed economy, it is the prerogative of the men to control its distribution throughout the band. And this pivotal position of meat distribution gives men, if not actual power, then certainly social prestige within the community. In societies based on a deep commitment to reciprocity and fair dealing, the giving of meat, both within bands and between them, inevitably confers on men a status that is simply not available to women. Meat opens to men important lines of communication and obligation.

Using just this one aspect of life in non-agricultural people, it is possible to predict the relative social positions of men and women in a particular society if one knows the style of their economy. The equation is simple: the more important meat is in their life, the greater relative dominance will the men command. For instance, in the Hadza in Tanzania and the Paliyans of southwest India the economy is very loosely organised with males and females more or less fending for themselves. Hadza men usually eat on the spot whatever meat they catch, and take back to the camp any surplus they may have. There is very little social or political discrimination between men and women in these people. By contrast, when meat becomes an important element within a more closely organised economic system so that rules exist for its distribution, then men already begin to swing the levers of power. This is the situation that exists among most gatherer-hunters, such as the !Kung, and it applies whether or not meat forms a majority of the diet. The status of women sinks further as we go into latitudes which enforce economies based more and more on meat. The Eskimos, in which virtually all food collection is carried out by the men, epitomise male domination of women. The women operate in a domestic world in which their social and sexual lives are totally dominated by the wishes of the men.

188

The differences in social structure that arise in different economic circumstances illustrate a degree of flexibility in the relationship between the sexes. But it is interesting that women's social standing is roughly equal to men's only when society itself is not formalised around rules for distributing meat. Even in the !Kung where the men provide perhaps one-third of the communities' food, their access to the important medium of exchange – meat – immediately gives them a social standing that their women do not have. It is a structured social organisation such as the !Kung's that we see as having been important in the evolution of *Homo sapiens*.

To suggest that the pressures of natural selection over the past two million years have made men in some way *superior* to women would be ludicrous. No one would dispute that on average men are bigger and stronger than their mates. Perhaps too our evolutionary career gave males a sharper visuospatial ability adapted for their quest for meat whereas women are more accomplished verbally, a talent well suited to the early education of their children: there is *some* psychological evidence to support this view. But no one has any basis for arguing that men are more intelligent than women.

In contemporary technological society, as in non-agricultural people, male dominance appears to be linked with control of valued resources. It is impossible not to be simplistic when commenting on the infinitely complex circumstances of affluent society, but it is feasible to view the equivalent valued resource as being access to money. The man who earns a large salary commands a certain respect among his fellows, and he may demonstrate his success by conspicuous consumption of impressive goods. Moreover, the executive's wife who elects to stay at home and leaf through the colour supplements deciding what new consumer goods she will try to persuade her successful husband to buy is more like an Eskimo than a !Kung. She accrues a certain status through being the mate of a successful man, of course, but her supreme dependence on him makes her so much easier to manipulate socially than a wife who works.

These days more and more women are breaking through tradition and are going out to work. And now that milk substitutes allow anyone to feed a baby, males' useless vestigial nipples no longer provide a biological reason why a woman should be largely responsible for child care, a factor that has usually fragmented female careers so that they could not compete with men on equal terms. All that stands in the way now of equal opportunity and equal achievement for men and women in social, economic, and political arenas is social tradition.

Or is it? If our ancestors pursued a lifestyle that was so successful partly because of the division of labour we have described, then perhaps we should expect that natural selection would have favoured the development of certain psychological attributes that would make men and women most efficient at the different jobs they performed. We see no way of denying that male *Homo sapiens* may have evolved a psychological approach to life that involves a drive for achievement within a framework of compatriatism with his male colleagues. And it would be surprising if evolution did not favour the survival of women whose skill and commitment to motherhood was of the utmost. These qualities in men and women would have contributed to individuals' genetic success and therefore to the ascendancy of the species as a whole. What we see in society today may be a cultural pattern woven upon the threads of basic biological imperatives.

This, of course, is an essentially unprovable proposition. In which case we perhaps should look to another aspect of culture for a solution – justice. Society is riddled with laws and prejudice that assume the inferiority of women. No such assumption is justified. And as we are a thoroughly cultural animal capable of ordering society as we choose, the only just way to proceed is to assume that men and women can achieve equally in social, economic, and political spheres. It can be done. But, as long-established social traditions have very deep roots, the Women's Liberation movement faces an enormous task in removing the expression of those traditions and replacing them with a set of values based on justice.

CHAPTER 11

Hunter to farmer:
giant leap or fatal step?

Anyone who saw Professor Alan Walker on the morning of 17 July 1975 might have thought he was crazy. Walker was in the hominid room at the National Museum in Nairobi and he had in front of him a one-and-a-half-million-year-old fossil cranium, probably the oldest, and certainly the most complete, specimen of *Homo erectus* yet to be discovered. He was poised over the priceless object with a cold chisel in one hand and a heavy duty hammer in the other. He turned the cranium upside down and placed it on a sand base; passed the cold chisel through the channel (the foramen magnum) that was once the exit for the ancient hominid's spinal cord; slotted the sharp point into a small hole that he had just drilled in the solid rock that now filled the space that once housed the hominid's brain; and then he struck the chisel with the hammer – very hard. It was a terrifying moment.

Nothing happened. So he hit it again. Still the ancient cranium remained intact. Walker persisted in his determined assault on the fossil for more than half an hour, until finally it split cleanly into two pieces, the fracture line snaking right across the cranium just behind where the ears would have been.

Walker wasn't engaged in palaeontological vandalism, nor some macabre rite. He had the job of cleaning the specimen, and breaking it apart was the only way of getting at the hard rock inside the cranium so that he could grind it away. Before attacking the fossil with hammer and chisel he had taken the precaution of encasing it in layer upon layer of paper and paste, exactly in the manner of making papier maché masks. The hard paper casing protected the petrified cranium, and it would have held the pieces together had the fossil shattered when Walker slammed the chisel with the hammer. Fortunately, it didn't shatter.

The whole episode was a remarkable piece of work, and it gave us a beautiful *Homo erectus* skull, the owner of which had lived – and died – by the shores of Lake Turkana at a time when our ancestors had reached a point in their evolutionary career at which they were ready to explore new lands outside the continent of Africa. As the direct descendant of *Homo habilis*, *Homo erectus* had built up an evolutionary momentum that was to propel it inexorably towards, first, primitive *Homo sapiens*, and ultimately to modern man. The emergence of the basic grade of *Homo sapiens* probably happened around half a million years ago, perhaps in Africa, or in Eurasia, perhaps in many different places at about the same time. The complex interaction of physical and intellectual capabilities within a self-created framework of culture probably operated on many populations of *Homo erectus*, urging them to the *sapiens* state. Not a divine guiding hand, but an inexorable biological progression.

The moment of birth of truly modern man is more difficult to pin down, and indeed the question of what exactly we mean by the term is a nice philosophical point. By convention, however, a group of 40,000-year-old skeletons, found in 1896 by workmen at the Cro-Magnon rock shelter in the limestone cliffs of Les Eyzies some 300 miles southwest of Paris, are the first examples of modern humans. But one could argue that it was only when agriculture was becoming well established that modern man truly arrived on the scene.

When, for whatever reason, bands of *Homo erectus* embarked on their journeys into Europe and Asia a million or more years ago, they had to ply their trade as gatherers and hunters in lands that were blessed with climates that were less kind, less certain, than in their tropical birthplace. For the first time our ancestors had to face marked changes of the seasons, with short but productive summers separated by long winters of scarcity. Ice ages closed in and retreated with uncertain rhythm, often making much of northern Eurasia totally uninhabitable. In case the perspective of this vital period of human evolution should be distorted, it should be emphasised that even during the balmier non-glacial intervals there were probably between five and 10 times as many hominids living in Africa as there were in Eurasia. And when the ice had closed in, probably no more than one in 20 of the world's hominid population was Eurasian. The widely popular idea that Europe was the focus of this phase of evolution – the transition from *erectus* to *sapiens* – is therefore unlikely to be correct.

The basic human qualities of opportunism and adaptability must have been essential to our ancestors as they moved into cooler lands of changing seasons, but their (undoubted) brown skins would have caused a problem. Not only was a layer of protective pigment no longer necessary against the penetrating rays of the absent tropical sun, but the pigment would also have prevented the more modest temperate sun from synthesising vitamin D in the skin. The biological answer was simple: much of the skin pigmentation was lost in people living in these cooler climes. (It is perhaps ironic that many pale-skinned westerners now spend a great deal of money on holidays each year encouraging what little pigmentation they do have to expose itself!)

The period between one million years before the present and, say, 10,000 years ago was truly remarkable in the evolutionary history of humankind. Our African cousins the australopithecines fell into extinction at the beginning of the period, perhaps succumbing to the ever-tightening grip of competition for resources exerted by *Homo erectus* on the one hand and the increasingly successful terrestrial monkeys such as baboons on the other. And the enormous evolutionary potential pent up in the populations of *Homo erectus* throughout Africa, Europe, and Asia generated a dynamic state of biological development, a matrix for progression towards the *sapiens* state, but with inevitable geographical variations. Some survived. Some did not.

Probably the most famous evolutionary failure of this period is the Neanderthal race, a stocky, beetle-browed people whose exaggerated physical features seem to have been adaptations to the extreme conditions of ice-shrouded northern Europe 100,000 years ago. The heavy features of the Neanderthal race are not entirely confined to the north, however, glimpses of them appearing further south in Europe and in the Middle East. But the degree of elaboration of such features, which represents the 'classic' Neanderthal, is strictly a northern phenomenon. And it is a phenomenon that appears to have blundered up an evolutionary blind alley. From 35,000 years ago there is no more recent sign of the Neanderthal race. They vanished. Perhaps they could not adapt to the warmer conditions that were easing their way into the continent. Perhaps they were unable to compete for resources with the more mainstream *Homo sapiens*. It is possible, however, that they may not have suffered complete biological oblivion: they may instead have interbred with the mainstream – the Neanderthals were, after all, just another brand, albeit an extreme one, of *Homo sapiens*. The genes of Neanderthalers may be surviving in us all today.

To view the birth of either basic *Homo sapiens* or that of fully modern man as a single geographical focus from which the new improved brand of humanity dispersed to overwhelm all existing populations before them is to impose too narrow a perspective on this unusual evolutionary stage of mankind. The perspective instead should be one of a dynamic interacting matrix of emerging populations, some of which would undoubtedly have been more successful than others, some, such as the Neanderthal race, may have wandered far from the mainstream, but most would have remained biologically compatible, the total effect being a broad thrust towards evolutionary advancement.

As millennium upon millennium passed by, our brains got a little bigger, our wits a little sharper, and – most important of all – our social and cultural fabric grew more elaborate and richly patterned: until, 10,000 years ago, we stood at the threshold of a revolution that was to transform the world. Following the final (or is it merely temporary?) retreat of the ice sheet around 12,000 years ago, the agricultural revolution began, slowly at first, but with an inexorably growing momentum. The switch from a gathering and hunting economy to one based on agriculture was an essential basis for the subsequent industrial and technological revolutions. It was also the spark that ignited the population explosion: because of the distribution of resources, a gathering and hunting way of life allows for a maximum world population of around 20 to 30 million; the greater concentrations of food produced by organised agriculture multiply that figure many times – it now stands in excess of 4000 million, and is still rising fast.

Why agriculture began when it did is something of a mystery. The retreat of the ice sheet may have been an important factor. In any case, the human race was clearly ready for it because many different groups of people embarked on this novel way of life at about the same time: Meso-America, the southeast edge of the Mediterranean (the so-called fertile crescent), and parts of southern Asia, all have archaeological evidence of primitive agriculture beginning around 10,000 years ago. These people cannot have been in touch with each other; they were simply too far apart. No, the practice of agriculture was invented simultaneously in many different places and it diffused out from these foci to revolutionise life in much of the rest of the world in less than 400 generations. Such is the power of cultural evolution.

The human evolutionary career has produced a highly unusual animal: an animal which has an ability arbitrarily to manipulate the environment unequalled by any other; an animal whose life is centred on the products of its own inventive mind; and an animal which, like no other, apparently has a penchant for waging organised murder on its own kind – warfare is totally foreign to the rest of the animal kingdom. Why?

Before we can start to try to answer this question it has to be admitted that war is an outrageously successful activity. History demonstrates this over and over again. In a world dominated by material possessions – whether of goods, land, or natural resources – a population of people may win for themselves enormous advantage through military victory over another group: the benefits gained must, of course, outweigh the costs of combat (time, resources, and lives). A materially based world undoubtedly provides a favourable environment in which warfare can flourish. And it has flourished more and more with the steady rise in the complexity of social structure. As American anthropologist Marshall Sahlins points out, 'war increases in intensity, bloodiness and duration . . . through the evolution of culture, reaching its culmination in modern civilisation.'

In many ways, the spectacular success of modern civilisation, of superstates, can be seen as a *product* of military manoeuvres. It is therefore ironic that the technology born of the super-states has now created an environment that is *unfavourable* to the practice of war. The invention and elaboration of sophisticated nuclear weapons means that in all-out war *there can be no winners*. In a nuclear holocaust *everybody loses*. We should need no reminder that the superstates' nuclear arsenals house the means of global destruction many times over. An extra-terrestrial observer might be intrigued to see how the earth creatures might adapt their political behaviour to this crucial environmental change. Here on *terra firma* we have a distinct personal interest in the outcome!

This lethal change in the environment in which belligerence is practised means that now more than ever before we need urgently to understand what it is that drives nations into combat with each other. The roots of human warfare have been under the scrutiny of political scientists and sociologists ever since the disciplines were invented. Biologists too have a legitimate interest in the matter because the behaviour of man may be viewed in relation to the rest of the animal kingdom. And for palaeoanthropologists the interest is paramount, because through learning about the evolutionary forces that shaped us into the creature that we are we may begin to have an insight into the nature of humankind. Is it in the nature of humans to wage war on each other? This is the key question.

Freud had an answer. In 1930 he wrote that 'Men were not gentle, friendly creatures wishing for love. . . . A powerful measure of desire for aggression has to be reckoned as a part of their instinctual endowments.' *Instinctual*? Yes, Freud is quite certain. 'Who has the courage to dispute it in the face of all the evidence in his own life and in history?' he asks. 'Anyone who calls to mind the atrocities of the early migrations, of the invasions by the Huns or by the so-called Mongols under Genghis Khan and Tamerlane, of the sack of Jerusalem by the pious crusaders, even indeed of the horrors of the last world war, will have to bow his head before the truth of this view of man.'

Earlier, in 1918, when the carnage of World War I was still fresh in his memory, Freud wrote: 'The very emphasis on the Commandment, Thou shall not kill, makes it certain that we are descended from an endlessly long chain of generations of murderers whose love of murder was in their blood as it is perhaps in our own.' Konrad Lorenz, who in 1974 received a share of the Nobel Prize for his part in pioneering the study of animal behaviour, was intrigued by what Freud had to say about human aggression. Lorenz was particularly interested in the nature of instincts, and in his famous book *On Aggression* he referred to his fellow Austrian, saying, 'It was Freud who first pointed out the essential spontaneity of instincts . . .'. And Lorenz warns that 'it is the spontaneity of the [aggressive] instinct that makes it so dangerous.'

Socially sanctioned murder – under the name of war – is, according to these views, the result of a basic drive within us, an instinct from which we can have no escape. Moreover, the instinct – Lorenz calls it aggression – is particularly dangerous because if it is not expressed following appropriate cues, it will erupt spontaneously. He comments of aggression in animals generally that 'There are few instinctive behaviours in which threshold-lowering and appetitive behaviour are so strongly marked . . .'. Referring specifically to human problems he says 'I believe . . . that present-day civilised man suffers from insufficient discharge of his aggressive drive. It is more than probable that the evil effects of the human aggressive drives, explained by Sigmund Freud as the results of a special death wish, simply derive from the fact that in prehistoric times intra-specific selection bred into man a measure of aggressive drive for which in the social order of today he finds no adequate outlet.'

Meanwhile, Raymond Dart, discoverer of the first australo-pithecine, was being greatly impressed by what he saw as undeniable evidence of early humans' carnivorous and even cannibalistic tendencies. In an essay entitled 'The predatory transition from ape to man', Dart wrote of our ancestors' proclivity for killing: 'On this thesis man's predecessors differed from living apes in being confirmed killers: carnivorous creatures that seized living quarries by violence, battered them to death, tore apart their broken bodies, dismembered them limb from limb, slaking their ravenous thirst with the hot blood of victims and greedily devouring livid writhing flesh'. This graphic fantasy becomes even more lurid when he suggests that in spite of our ancestors' small brains, 'this microcephalic mental equipment was demonstrably more than adequate for the crude, omnivorous, cannibalistic, bone-club wielding, jaw-bone-cleaving Samsonian phase of human emergence. . . . The loathsome cruelty of mankind to man forms one of his inescapable, characteristic and differentiative features; it is explicable only in terms of his carnivorous and cannibalistic origin . . .'. Dart was basing his pronouncements on damage he saw in fossilised baboon and early hominid skulls, damage that he considered must have been the death blows from weapon-wielding pre-humans.

Lorenz was not slow to adduce the so-called evidence of pre-historic carnage to support his thesis of man's innate blood lust. He tells us that once our ancestors learned the trick of making stone weapons there was no natural inhibition preventing their being put to deadly purpose, man against man. 'There is evidence,' he says, 'that the first inventors of pebble tools, the African australopithecines, promptly used their new weapons to kill not only game, but fellow members of their own species as well. Peking Man, the Prometheus who learned to preserve fire, used it to roast his brothers: beside the first traces of the regular use of fire lie the mutilated and roasted bones of *Sinanthropus pekinensis* (*Homo erectus*) himself.'

This compelling view of the bloodthirsty side of human nature which for a long time was promulgated without particularly dramatic impact by eminent scientists was finally propelled into full public prominence by the skilfully evocative pen of Robert Ardrey. Author of *African Genesis*, *The Territorial Imperative*, *The Social Contract*, and, more recently, *The Hunting Hypothesis*, Ardrey was born in Chicago's south side in 1908. As a youth in Chicago he took a course in social science under Professor Ogburn. He once lectured in anthrolopogy at one of the booths at Chicago's World Fair, and then abandoned what can at most be described as a brief acquaintance with the science to pursue a career as a playwright. While in Hollywood he wrote the script of an epic military film, *Khartoum*. With an enforced interval of the second world war, he continued his script-writing career until 1954. Then, out of work, his new Broadway play a disaster, he was asked to do a series of articles for *The Reporter* on Africa, particularly on the Mau Mau uprising in Kenya.

It was while Ardrey was in Africa carrying out his commission that he met, almost by accident, Raymond Dart. It was a meeting that was to change Ardrey's life. As with Saul on the road to Damascus, Ardrey suddenly saw clearly the goal to which he should direct his energies. Dart's fervent enthusiasm, the ancient crushed bones, the jagged stones, all combined to fire Ardrey with a new purpose: of revealing to humanity the ugly truth about itself, that deeply rooted in our being is a lust for flesh, for killing, that *must* be sated.

Ardrey may have made little impact on Broadway, but there can be no denying that his impact on a global audience has excelled that of the world's greatest playwrights. He wove together fact and opinion on a loom of brilliant persuasive prose. He told a story that was more gripping than most inventions of fiction. But most important, the story he told was one that people wanted to hear: that the bloody carnage that stains much of recorded human history is an inescapable expression of basic human nature. He says that 'The human being in the most fundamental aspect of his soul and body is nature's last if temporary word on the subject of the armed predator. And human history must be read in these terms.' 'Man,' we are told, 'is a predator whose natural instinct is to kill with a weapon.'

Viewing the emergence of an ability to make stone tools – Ardrey prefers to call them weapons – deep in our history, Ardrey has this to say: 'Far from the truth lay the antique assumption that man fathered the weapon. The weapon, instead, fathered the man. The mightiest of predators had come about as the logical conclusion to an evolutionary transition. With his big brain and his stone axes, man annihilated a predecessor who fought only with bones. And if all human history from that date has turned on superior weapons, then it is for very sound reasons. It is for genetic necessity. We design and compete with our weapons as birds build distinctive nests.' In other words, according to Ardrey, human evolution has been driven by an engine of warfare: the band whose genetic inheritance endowed the means to make and exploit to devastating effect superior weapons of war survived and thrived at the expense of more peaceable people.

The formula is simple and the appeal direct. In a perverse sense it is comforting to be told that war is beyond the control of even reasonable men. If bloody conflict really is written indelibly in our genes, then there is no point in agonising over it. If war is inevitable because of our very nature, we need not trouble our conscience about it. We are absolved from responsibility. We need nurse no guilt.

The arguments that Ardrey and his like adduce to support the human innate depravity come from worlds living and worlds long dead. On the one hand there is the animal instinct of aggression, an instinct that is often closely linked with territoriality. And on the other are shattered fossils, putative signs of prehistoric violence.

Anyone who has even a cursory glance at various corners of the animal kingdom very soon sees that aggression – or at least the kind of behaviour that is *called* aggression – is a very common everyday experience. Animals fight over food, over space, and over mates. It is this fighting that ethologists and others have termed aggression. In fact in their aggressive encounters animals rarely come to blows: only very, very rarely do they deploy their claws, teeth, and whatever other dangerous equipment they might have in a way that suggests that they mean business.

Social dynamics of animal communities, and the ecological conditions in which they operate, demand that individuals must always maximise their chances of access to important resources such as food and mates (territoriality is an integral part of both these factors). But the reason that animals do not engage in deadly serious combat over these resources is that fighting is an expensive business: it is expensive in time, energy, and life itself. So, when two males contest the right to, say, a female, they engage in ritual battle in which they both adhere to a set of unwritten rules. The combatants both try to bluff the other into believing that they are facing superior individuals and that serious trouble might follow unless submission comes pretty soon. Sure enough, one of the contestants will recognise the apparently superior aggressive signals of the other and will throw in the towel by appropriate submissive postures. Usually, such submission puts the loser in an extremely vulnerable position, perhaps exposing his throat to the jaws of his opponent. But, instead of pushing his advantage to the full, the victor acknowledges the signal and the fight is over.

Why does the winner not finish the job and sink his fangs into the throat of the vanquished? Because if such combats really were pushed to the death the loser would not simply offer his opponent an easy victory: he would fight frantically to the last breath, and in the end the winner might have to pay dearly for his conquest. It is therefore in every individual's interest to settle disputes over food or mates in short stylised confrontations rather than in costly battles. In the animal kingdom individuals are in business not to dispose of their fellow competitors but to outbluff them.

Proponents of human aggression often claim that modern man lacks the instinctive responses that in other animals prevent combat escalating to lethal levels. Some people, such as Desmond Morris of *Naked Ape* fame, suggest that the sudden acquisition of highly effective stone 'weapons' was too great a change in the delicate balance of individual conflict for subtle signals to be able to divert disaster. If, for the sake of argument, early hominids did employ their stone artefacts as weapons against each other, then there is absolutely no reason to believe that, over the long evolutionary period we are contemplating, the basic rules of biology did not apply. In this scenario there may well have been an initial distortion of inter-personal conflict, but very soon the rules of biology would have reasserted themselves. Indeed, it would have been evolutionary suicide if our ancestors had cast aside the laws of conflict and converted every dispute into a potential murder. The costs would have been very high, not only for the victims, but for the murderers too. Such an animal would not have survived.

In *On Aggression* Lorenz offers a suggestion for the lack of inhibitions against murder in humans by assuming that such events would be quick single-blow affairs: 'In human evolution, no inhibitory mechanisms preventing sudden manslaughter were necessary, because quick killing was impossible anyhow . . . until, all of a sudden, the invention of artificial weapons upset the equilibrium of killing potential and social inhibitions. When it did, man's position was very nearly that of a dove which, by some unnatural trick of nature, had suddenly acquired the beak of a raven. One shudders at the thought of a creature as irascible as all pre-human primates are, swinging a well-sharpened handaxe.' One wonders why Lorenz describes our primate cousins the chimp and the gorilla as 'irascible'. During explorers' first encounters with these creatures they may well have appeared so. But the now well-established studies of chimps and gorillas (by Jane Goodall, Diane Fossey, and their many colleagues) show that these apes are not irascible creatures by any means. And even if they were, would it necessarily be logical to conclude that the early hominids were equally bad-tempered?

One of the major flaws in the argument about aggression is the assumption that it is an unwavering instinct. Not only do the rules of biology operate to minimise the intensity of aggressive encounters, but the behaviour itself is very responsive to environmental conditions. A species may be outrageously territorial in one kind of environment, but not at all in another. For instance, when the ayu, a salmonid fish, is in shallow brooks individuals aggressively defend small territories against intrusion by others. And yet when they are in deep pools they move in large shoals. In populations of red grouse individuals assiduously mark out territories where they feed and breed. By a simple experiment a British biologist showed that the size of the territory depends on the density of protein available and not on an immutable drive wired into the bird's brain. He merely fertilised a large area of heathland where the birds staked their claims, and, 15 months later, when the vegetation was much more lush than usual, there were many more birds in the area, and each occupied smaller than 'normal' territories. Vervet monkeys living in cramped marginal environments can become aggressively possessive over resources, whereas in luxuriant surroundings their sharp aggression and territoriality diminishes.

Male chimpanzees sometimes indulge in mild disputes over the favours of females 'on heat'. But if there are a lot of male contenders the aggression diminishes. The reason appears to be that when there are lots of potential mates around, any individual who spends much of his time fighting with others is wasting opportunities for copulation: his other competitors would meanwhile be lavishing their attentions over the disputed female! And there is a magnificent, but recently injured, male red deer living on the Isle of Rum off the west coast of Scotland that gives an elegant demonstration of the plasticity of so-called aggression. The large stags control harems of perhaps six females which they win in trials of strength with other stags. Harem owners periodically announce their ownership by thunderous roars that echo around the hills of the beautiful island, an impressive display of ownership aggression. One stag snapped a tendon in one of its legs in an accident, and this meant that it could no longer engage in powerful head-to-head pushing contests with other males. Undismayed, however, the stag continued its impressive roaring, but only when it was without a harem. When he managed to assemble a few mates in his ownership he kept unusually quiet. And this made very good sense because to have advertised his harem abroad would have invited the interest of other males, males against whom he would have been unable to compete in the trial of strength.

There are now many examples that illustrate the sensitive responsiveness of aggression and territoriality to prevailing environmental (and personal) conditions. These instincts are by no means as rigid as once was imagined. And the appetite for aggression most certainly cannot be thought of as a steadily rising head of pressure that must be vented lest it explode dangerously. To talk of the appetite for aggression – particularly territorial aggression as Ardrey does – is to promulgate a mechanistic view of animal behaviour that bears very little relation to the observed facts.

Although we can never be certain, it seems entirely reasonable that our early ancestors occasionally had reason for dispute, both between individuals and between bands. Such disputes would have been heightened at times of scarcity of important resources (women or food). To suggest anything else would be to plead that hominids did not belong to the animal kingdom. But to suggest that our ancestors were in a continual state of war makes no biological sense. Lorenz talks of our ancestors' 'hostile neighbours' for no better reason than that it suits his story.

The greatest fallacy, though, in the thesis of human carnage is to equate animal aggression with organised warfare. The stylised behaviours that animals perform during confrontations over resources are very much individualistic affairs. A nation going to war has very little to do with individual psychology or behaviour. It is a political response to a political opportunity or threat. There is no biological connection between an army training for battle and a wide-mouthed yawn threat that a male baboon gives to a temporary competitor. It is for this reason that one has to be very careful in the use of the word 'aggression'. When the same word is used to describe the political posturings of a nation poised on the brink of war and the stylised posturings of two males in contest over a female, it is very easy to be seduced into an argument that both sets of behaviours have the same root. They do not.

If our ancestors had been particularly war-like with each other, how would we know? Dart has an answer, and he saw it in the damage suffered by the hominid fossils recovered from the South African caves. For instance, the first hominid to be found in the region, the *Australopithecus africanus* child from Taung, has a number of fracture regions around the face and what remained of the cranium. Dart suggested that the child had suffered 'A lateral blow on the left fronto-temporal region of the skull'. Another fragmented skull, this time an adult from Sterkfontein, which lacked the lower jaw and much of the face was said by Dart to have met its end through 'a lateral blow on the left tempero-parietal region of the skull'. Another victim had been killed, according to Dart, by 'a vertical blow just behind and to the right of the bregma with a double-headed object'. A third individual from the same cave showed evidence, again according to Dart, of 'A vertical blow slightly to the left of the mid-parietal region with a bludgeon'.

Part of the back of a cranium found at the Makapansgat cave was said to be the result of 'a severing transverse blow with a bludgeon on the vertex and tearing apart of the front and back halves of the broken skull'. From the same cave came a damaged lower jaw of an adolescent (about 12 years old) who, Dart tells us, had met a violent death: 'The fractures exhibited by the mandible (lower jaw) show that the violence, which probably occurred in fatal combat, was a localised crushing impact received by the face slightly to the left of the midline in the incisor region, and administered presumably by a bludgeon. The result of that decisive blow, as far as the mandible is concerned, was that the four permanent incisors (and perhaps the left second deciduous molar) were sprung from their sockets and the bone was shattered.'

The list of putative prehistoric violence is long, and Dart's writings are replete with detailed considerations of extent of damage and the 'probable causes'. It was a list that greatly impressed Ardrey in his investigations of our 'blood bespattered' past. Bob Brain, however, is less impressed. He has been making extensive studies of the means by which bones accumulated in the caves and what happened to the bones once they arrived there. It was Brain, remember, who effectively disposed of Dart's notion of an osteodontokeratic culture, showing that the unusual selection of bones in the caves was the result mainly of the resistance of different types of bones to the gnawing attentions of scavengers.

Once a skull, for example, has tumbled into a cave it is steadily covered by dust and stones, and, of course, other bones. As the weight of dust and stones covering the skull gets greater the cranium will begin to be distorted, unless, that is, the empty cavity rapidly becomes filled with hardening dust that gives the bone some support. Most skulls are distorted, and many may have rocks thrust into them by the sheer burden of the overlying deposits. If the cave breaks down and the deposits are exposed to erosion the now-fossilised bone is once again vulnerable to damage.

In a recent paper on his findings Brain says of the cave accumulations that 'once buried within the deposit, [the bones] may suffer the most remarkable distortions, fractures and dislocations as a result of pressure and movements of the sediment. The commonest kind of distortion suffered by bones preserved in a cave breccia results from the overburden pressure and takes the form of flattening. Skulls whose endocranial cavities have not been filled with matrix are typically flattened as if they have been run over by a steamroller, while those that have been partly or wholly filled with matrix will be affected in accordance with the nature of the filling, the surrounding breccia and the force exerted.' Those seeking signs of prehistoric violence beware!

A cautious man, Brain says 'In view of this complication, it is frequently difficult to decide whether a fossil has suffered its damage before or after its burial in a limestone cave deposit . . . it may never be possible to say with certainty what the cause of the observed damage might have been.' He acknowledges that 'Anthropologists who accept Dart's osteodontokeratic culture will doubtless be inclined to agree that the Makapansgat remains may be taken as evidence of interpersonal violence and cannibalism.' If this is indeed the case then the Pleistocene must have been a hazardous time to have been alive, for virtually 100 per cent of the hominid skulls recovered in the South African cave deposits show signs of damage that have been interpreted by at least one authority as being evidence of a violent death! Although Brain is able to offer alternative explanations involving the nature of fossilisation in cave breccia in all these cases, he prefers not to be dogmatic: 'the question of interpersonal violence among the australopithecines must remain an open one'. Holes in the head made by the pressure of rocks in cave deposits make less dramatic stories than those inflicted by stone weapons wielded by the hands of blood-lusting hominids. But they may be nearer the truth.

Less equivocal evidence of the involvement of 'human' hands in the damage to ancient skulls comes from the Choukoutien caves near to Peking. There, around half a million years ago, groups of *Homo erectus* sat with the skulls of some of their fellow creatures before them and carefully widened the foramen magnum so as to get at the brain. Probably they did it so that they could eat the brains of the dead. These are the bones to which Lorenz refers when he suggests that our ancestors 'used their new weapons to kill not only game, but fellow members of their own species as well'. The suggestion of cannibalism seems incontrovertible. But the crucial question to ask is, what was the context of the cannibalism? Were the people whose brains were eaten the victims of a bloody massacre? Or were they the relatives of the people who consumed them, an act performed to seek a continuation of the dead through the living. Both forms of cannibalism have been indulged in by technologically primitive people until recently. As evidence of ancient cannibalism is not confined to China but occurs in other parts of Asia and in Europe too, we have to consider whether our ancestors made a habit of killing and eating their enemies in acts of war or ate their fellows as a mark of love and respect.

Eating the bodies, or parts of the bodies, of dead relatives or members of the tribe – known as endocannibalism – has been practised in many parts of the world. In the Dieri, an aboriginal people of southeastern Australia, when someone died an old man who was a relative cut all the fat from the face, belly, arms, and legs, and then handed it round to eat. These people believe that a person's fat contains unusual powers which may be acquired by those who eat it. The Dieri also believe that by eating the fat from their dead fellow they are absorbing his personality and soul, qualities which are therefore preserved within the tribe.

In some South American tribes, such as the Amahuaca, Tucana, Waika, Pakidai, and Jumano, the people say that the home of the soul is in the bones, not the fat. These people burn their dead and then mix the ashes of the bones with their drinks. In this way the soul takes up residence in the people who drink the mixture. In another tribe, the Chiribichi, the dead are slowly roasted and the melting fat is collected and drunk. There are many different ways in which endocannibalism is practised, and the rites and customs surrounding it may be very complicated. In all cases, however, the purpose is the same: a continuity with the dead.

Exocannibalism, the opposite of endocannibalism, also takes many different forms, and in recent times it was the more common of the two. When the Sumo, a South American Indian tribe, had slain an enemy they chopped him up and ate him, both as an insult and also to prevent him taking any kind of revenge. The Parintintin, another South American people, were more selective in their exocannibalism. They cut out the eyes, tongue, and muscles from the arms and legs of their dead enemies and eat them, the idea being to prevent the people from seeing, talking, walking, and shooting again. A man slain by the Cubeo tribe would have his penis eaten by the wife of the chief at the end of a meal. The belief was that it increased her fertility.

Exocannibalism, like the more peaceable form, is therefore not simply a way of supplementing the diet: it is part of the cultural framework of the society. In fact, human bodies are not an especially plentiful source of protein. If, for instance, an averaged sized group of gatherer-hunters decided that they would include humans on their menu so as to provide them with half of their protein needs then they would have to slaughter a man every other day. Theoretically feasible perhaps, but hardly a way of providing a viable lifestyle!

In an attempt to decide which of the two forms of cannibalism is the more primitive, German anthropologist Hermann Helmuth did an analysis of South American tribes. According to the *Handbook of South American Indians* there were 16 recorded peoples who ate their friends against 38 who ate their enemies. The point of interest, however, is whether there was any difference between pure gatherer-hunters and agriculturalists. There was. Two-thirds of gatherer-hunters who indulged in cannibalism were endocannibals. By contrast 32 out of 34 cannibalistic agricultural people were exocannibals. If exocannibalism can be seen as more 'aggressive' than endocannibalism, then we can suggest that an agricultural economy leads people to be more aggressive. A gathering and hunting economy, the most primitive economy known to man, is more closely characterised by peaceable cannibalism. We can never be sure, of course, but it does perhaps suggest that the signs of cannibalism left behind by our gathering and hunting ancestors are more likely to have been the result of endocannibalism than of exocannibalism.

Aside from prehistoric signs of ancient 'death blows' and cannibalism, we can search for more indirect signs of violence our ancestors may have directed at each other. Recorded history begins only a few thousand years ago with the invention of writing – the first examples are mainly of inventories of grain stocks and the like, but accounts of battles are prominent enough to suggest that warfare was more than an uncommon activity 5000 years ago. Instead of writing, we can look at prehistoric painting for clues of what our forebears did in their spare time.

Homo sapiens first took up the brush – or whatever implement they happened to use for decorating the walls of caves or rock shelters – around 25,000 years ago. The main preoccupation of rock artists was to depict the form of animals, presumably the prey of prehistoric gatherer-hunters. The artists may have painted animals simply for enjoyment, but a more generally accepted interpretation is that the pictures formed part of a ritual designed to increase the success of the hunt. Or perhaps they felt the need to appease the gods – or their own consciences – for their necessary destruction of life. Certainly, there are no serious representations in early art of the plant foods on which the artists undoubtedly depended for their existence. No magic is needed to collect nuts and roots, and, as they have no readily apparent vibrant spirit, the gods need not be appeased for their destruction.

Also absent, apart from very, very few examples, from prehistoric art prior to the agricultural revolution, is depiction of war. There are two possible explanations for this omission. First, that warfare *was* common before our ancestors began to abandon the gathering and hunting economy in favour of sedentary agriculture, but that for some reason – lack of interest or the fear of inviting bad luck perhaps – artists did not reproduce images of it. Or, second, that it *was not*. After the agricultural revolution, when we began to find other evidence of war, for example heavily fortified towns such as Jericho in what is now Israel, artists did paint pictures of battles, particularly ones in which they were victorious.

Once again we cannot *know* why pre-agricultural revolution art makes only scant reference to war, whereas art after the revolution does depict battles which we know from other archaeological evidence did take place. But one reasonable explanation must be that organised murder was a rare and unimportant element in the way of life.

In sifting through the evidence of armed strife in times past – whether it is in the area of animal aggression, shattered fossils, signs of cannibalism, or indirect clues such as art – there is nothing about which we can be absolutely *certain*, nothing from which we can deduce, yes we are descended from blood-lusting ancestors. What we must do is rationally balance a set of relative uncertainties and see which way the scales tip. If animal aggression cannot be seen as a sufficient drive for organised warfare, if skills of early hominids might have been shattered just as easily by their precarious journey into the fossil record as by a crushing blow from a bludgeon, if undeniable signs of cannibalism are marginally more likely to have been the result of care and respect rather than malevolence, and if records of battles are almost totally absent before the agricultural revolution, which way might we see the scales moving?

An objective assessment must surely admit that the weight of evidence is in favour of a relatively peaceable past. Without doubt there must have been *some* interpersonal violence: some of our ancestors must have met their end at the hands of their fellows. And we cannot deny that, occasionally, neighbouring groups may have become sufficiently hostile to each other to pitch themselves in deadly battle. But we can see no argument persuasive enough to suggest that warfare was a prominent feature of the past, that it was an important engine of human evolution.

But what of the wars of history? Do they tell us nothing of human nature? The fact that during the past 10 millennia war has become more and more important in the affairs of mankind undoubtedly tells us something. First, and most simplistic, it tells us what we know already, that man is *capable* of waging war, and very effectively so. But this does not mean that the specific activity of war is written into our genes, any more than the specific skill to play the game of football, the specific talent for making fine wine, or the specific inventiveness to design interplanetary rockets. We are an animal of many *potential* talents, and as we live in a world that is largely of our own making, we can decide in what direction we should channel those talents. As we've said, in a materialistic world, war is a highly successful way of gaining material advantage. It is therefore no surprise at all that war became such a popular way of getting what was wanted. But this is very different from saying that there is a biological instinct within us that drives us into battle, whether it is advantageous or not.

The turning point in our history came with the invention of an economy that allowed a previously nomadic people to live in large numbers in villages supported by an abundance of husbanded food (when the harvest was good). Gatherer-hunters generally live in small bands (around 25 people), move around regularly, and have no more possessions than they can carry on their back. They also limit their birth rate to once every four years so as to fit in with the mother's mobility in gathering food and the band's mobility in moving to new locations every few weeks. Once an abundant supply of food is available in a single place, these constraints are lifted. Birth rate can increase, the population therefore starts to grow, and people can accumulate possessions.

Incidentally, not all pre-agricultural revolution gatherer-hunters lived nomadic lives in search of widely distributed food. Some lived in areas of plenty and were able to build villages occupied by many hundreds of people. One such is the village of Lepenski Vir, perched on a terrace within the rugged Iron Gate gorges of the Danube. The people of Lepenski Vir lived well on the huge sturgeon and carp that are still thriving in the river. So abundant was their food source that the people were able to build large trapezoidal houses with paved floors and stone-lined hearths. Their settled village life allowed them to develop an elaborate material culture, one expression of which was carving fish-like faces on boulders, presumably part of a fish-like deity.

More modern examples of non-agricultural peoples exploiting plentiful food resources are the Indians of the northwest coast of North America and the Ainu who inhabit Hokkaido, an island north of the mainland of Japan. Hunting, and particularly fishing, were so productive that these people lived in large villages and evolved complex social structures not seen in nomadic gatherer-hunters. They had well-developed social stratification, at the head of which was a chief. They had a currency, and with that came wealth and prestige. And because of seasonality of food supply, they developed well-organised systems of food storage. The important point about this is that it is not agriculture itself that leads to village and town life and the material elaborations that go with it, but simply a plentiful and reliable supply of food in a single location. Gathering and hunting people have a plentiful supply of food, but they must move around to exploit it. They do not create an elaborate material culture or social stratification. And their only currency is the skill with which they seek out plant food and meat.

Although an agricultural community may prosper with their husbanded products of the land, they are at risk not only from a poor harvest, but also, and this is more important in the course of human history, from neighbouring communities who might covet their crops. As soon as people depend on anything so discrete as a standing crop, then there is advantage to be had in purloining one's neighbour's crops. Naturally, the community under threat will fight furiously to keep what is theirs because without the harvest they are lost: a new crop of food takes at least a year to be ready, especially if forest is to be cleared and the land prepared. For instance, the Yanomono Indians of southern Venezuela ('the fierce people') grow bananas, plantains and other food in a large garden close to their forest village. If they are forced to abandon their village following a raid from one of their aggressive neighbours, they must spend at least a year with an ally until they have prepared a new garden. Living with an ally is very burdensome because not only does it obligate them to support their ally in the future, but they also have to pay for their hospitality in women. Although Yanomono women are not treated well in humanitarian terms, they are a highly valued resource as they produce sons who can be warriors. To lose women even to allies is a terrible price to pay.

Possessions, whether of food or other valued materials, invite attempts to gain them by easy means. And as humans can be claimed to be neither inherently evil nor inherently good, but simply opportunistic, it is inevitable that some people will respond to such an invitation. And once the successful cycle of raiding begins it is very difficult to break. In an environment in which a particular form of behaviour is advantageous, that behaviour will persist. War is an advantageous pursuit in a material world. But it is a product of cultural invention, not a fundamental biological instinct.

When human communities become large, social stratification, including chiefs and leaders, appears inevitable, especially in societies placing particular emphasis on material wealth. Such a social structure is, of course, well suited – even essential – to waging organised conflict. Without a powerful leader it would be impossible to rouse the enthusiasm of the masses into an efficient army. It is perhaps ironic that the one characteristic that must have been vital in the evolution of a gathering-and-hunting economy in early humans – cooperation – should also be crucial to organised warfare!

In considering conflict between nations in the modern context, especially now that the advanced nuclear technology of war has made it a potentially lethal pursuit for all parties involved, and for non-involved bystanders too, it is perhaps irrelevant to scrutinise closely our biological history. We have done so at some length because of the prominence that such interpretations of human 'aggression' have received in recent years. Perhaps for the reason that it offered an easy absolution, the notion that the armed conflicts we see in the world today are the product of our biological evolutionary career has been eagerly assimilated. The notion, as we have shown, is in all probability a myth. But worse, it diverts the attention away from the more immediate pressing problems that face humanity and from the uniquely human social and political context in which wars are waged.

The prospects for most of the 4000 million people alive today are poverty, starvation, poor health, economic exploitation, and political impotence. Grim prospects indeed, and they get grimmer as relative and absolute deprivations deepen, while the dream of sampling just some of the wonders of modern civilisation fades inexorably away. War must be seen in the socio-economic context of the human condition. It is a political institution controlled by powerful people, whether it is waged between two neighbouring tribes, or two superstates. We should not be looking into the jaws of fighting baboons in the hope of finding the cause of war. We should be examining the structure and motivation of power politics.

In the new environment in which war would take place there can be but one excursion into global conflict. A nuclear holocaust could be the means of extinction of *Homo sapiens*. Perhaps this is inevitable. Perhaps when *Ramapithecus* stood upright all those millions of years ago it was setting off on a journey that ends in yet another evolutionary blind alley. Many species have faced the same fate. But in our case extinction would be entirely of our own making, the result of being intelligent enough to create the means of our own destruction but not rational enough to ensure that they are not used.

There is probably every reason to be pessimistic about the prospects of a third world war. A. J. P. Taylor, the famous British historian, sees it as almost unavoidable, given the politics that people seem to like to play. But perhaps it *is* possible to drive a wedge between our present behaviour and that bleak inevitability. That wedge can come from our study of human prehistory, a study that tells us of the common origin of humankind and of the basic characteristics – cooperation and sharing – that nurtured our long evolution. Biology shows us our common biological roots and teaches us that the gulfs that divide nations are cultural and political artifacts. Just as the human urge for cooperation has in the past been exploited in the politically motivated context of war, we can expect that the same urge could be harnessed in a politically motivated context of peace. It is the right political motivation that is needed. And the basis for that political motivation should be that not only do we all share a common origin, but, inescapably, we share the same destiny. It is a destiny that the human race is now capable of choosing.

Index

Index

INDEX

G/wi San, 83–4, 86

Hadar, 19, 52, 58, 60, 61, 64–5, 67, 75, 115, 163
Hadza people, 188
Hanuman langurs, 34, 178
Hearing, *see* Ears
Hill, Andrew, 64
Himalayas, 19
Hobbes, Thomas, 85, 86
Holloway, Ralph, 127ff, 154, 161, 162
Hominidae (human family), 27, 40
Hominids, 9–10, 16–17, 45–7, 53–4, 139
  ancestry, 24, 45–6
  appearance, 54, 60, 68
  bones, 45–6, 61–2, 65ff
  brain, 127ff, 161–3, 166, 167
  defined, 17
  demise, 46–7
  diet, 54, 71–3
  environment, 53, 57
  first, 27
  fossils, 45–6, 65
  language, 160–1, 166–7
  sex, 172, 181–2
  tools, 105–6, 163–6
Homo, 44, 51, 52, 54, 63–4, 66, 68, 70, 71–2, 106, 109, 113, 115, 123, 125–6, 128, 131, 132, 133, 163
  altruism, 123
  appearance, 68
  bones, 63–4, 66, 70–1
  brain, 128, 131, 132, 133, 162
  choice of mate, 184
  food, 71–2
  hunting, 115
  language, 153
  sex life, 180
  skull 1470, 51, 52
  story telling, 159
  tools, 72, 163
Homo erectus, 63, 66, 67, 77, 132, 139, 154, 162, 191–3, 198, 205–6
Homo habilis ('able man'), 10, 46, 47, 49, 56, 60, 63, 66, 67, 68, 77, 162, 181, 192
Homo sapiens, 39, 40–1, 46, 63, 67, 95, 108, 120, 133, 139, 154, 192, 193, 194

brain, 133
extinction, 212
Hrdy, Sarah Blaffer, 178
Humans, 29, 77, 116, 158, 169, 195, 209
  behaviour, 160
  brain, 130, 158
  culture, 39–40, 170
  evolution, 29–30, 35, 36–7, 40, 98, 169–70, 192–3
  intelligence, 128
  language, 158
Humphrey, Nicholas, 138, 147
Hungary, 27, 169
Hunting-and-gathering society, *see* Gathering-hunting society
Hunting, 88, 92–4, 104–5, 115ff, 186–7
  earliest, 116
  economics, 89, 186
  totem animal, 171
Hunting Hypothesis, The, 104, 198
Hyaena, 21, 48, 57, 63, 87, 92, 110, 111, 115

Imagery, 158
Importance of Being Earnest, The, 127
Incest, 88, 100–2, 171, 180
India, 27, 28, 38, 98, 178, 186, 188
Indians (of North America), 210
Infanticide, 96–7, 178
Intelligence, 128–9, 133ff
Isaac, Glynn, 111, 164

Jericho, 208
Johanson, Don, 43–4, 46, 60–2, 64–5, 66, 69
Johanson's knee, 69–70
Jones, Nicholas Blurton
  see Blurton Jones, Nicholas

Kalahari Desert, 81, 83
Karari technology, 76
KBS site, 22, 73–4, 76, 105, 115, 163
Kenya, 10, 15, 16, 17, 20, 27, 63, 98, 103
Kenyapithecus, 28
Kimeu, Kamoya, 11, 54
Kin selection, 122
Kinship, 82, 99–101, 122, 182
Koobi Fora, 9, 17, 19, 20, 39, 48,